THEO-COSMOS:

A
Scientific Description
Of The
Universe
From A
Theological Perspective
Secrets of The Universe
The Quest for The Knowledge
of The Universe

Rev. Byran C. Russell

KS
Kravitz & Sons
INNOVATORS IN PUBLISHING, MARKETING AND ADVERTISING

Kravitz and Sons LLC
1301 Farmville Blvd, Suite 104
Greenville, NC 27834

Published by Kravitz and Sons LLC.
ISBN: 979-8-89639-113-5 (sc)
ISBN: 979-8-89639-112-8 (e)

Library of Congress Control Number: 2025903999

OBJECTIVE

Many Scientists and Theologians have attempted to answer some of the Fundamental Questions of the Universe, yet there are now more questions than answers.

- 1. What is the cause of the Universe?
- 2. What is the purpose of the Universe?
- 3. If the Universe has a cause and purpose, What is its destiny?
- 4. Has the Mechanism of the Universe been adequately described?
- 5. Why is the Universe in a state of weightlessness?
- 6. Have the microphysical processes and entities lacking in Newton Mechanics been discovered and clearly described?
- 7. Will the Regular Sequence Stars deplete their nuclear fuel over a period of time and will the rotation of the Galaxy cause its constellation to disintegrate over a period of time?

These answers are herein provided, thus allowing the human eye to peer, for the first time, at the heart of the Cosmos.

Byran C. Russell

FOREWORD

There is a sense of satisfaction that the scientific deductions herein postulated against a theological background will enlighten our understanding of the Universe. The reviews of scientific theories and all other subjects herein considered are devoted to that end. The answers arising from these subjects are the answers to the fundamental questions of the Universe. No stone that would impede the progress to a better understanding of the Universe was left unturned. However, recognition of the Gap Theory will add much clarity to the discussion that follows

The Gap Theory distinguishes between the Dateless Past of Creation in Genesis chap.1:1-2 and the beginning of time with the creation of light, the first day of Genesis 1:3-2: 1-24. The Theory recognizes an immense gap of time between the Dateless Past of Genesis 1: 1-2 and the Six Day Creation beginning at Genesis 1:3-2:1-24.This gap of time could be billions of years. It is important to observe that the first day of Creation begins with an Earth already in existence. Also, on the first day the stars were already created but their lights were not shining on the Earth. God simply switched on the lights, particularly that of the Sun.

We have often overlooked Scriptures which evidently support the Gap Theory. We cite Job 38:4-7—"Where was thou when I laid the foundations of the Earth…When the morning stars sang together, and the sons of God shouted for joy?" This is the strongest evidence of Creation in the Dateless Past. It is against this background Theo-Cosmos must be viewed.

The Universe is God's greatest investment, the supreme Divine act ; and, like nothing else, has compelled our attention. Yet, behind all of this, there is a purpose, a purpose so profound that it cannot be fully realized in our life time.

Byran C. Russell.

TABLE OF CONTENTS

INTRODUCTION

Man finds himself living in an amazing Universe. He looks around him and discovers that every bit of it is real to the very atom. He wants to know the cause of the Universe, how and when it was formed, its purpose and destiny. He finds that there are laws of every kind everywhere. Above all, he finds that some of these laws are biological and moral in nature; and, therefore, place him under those restrictions. Man is thus impelled into the belief of a great Cause which he is trying to identify.

Man's quest for a fuller knowledge of the Universe is justified and is within the limits of his moral rights. The extraordinary significance of each of these answers in man's quest for knowledge implies that they could not be left to be determined by scientific means, nor by reason, nor by conscience. The CREATOR Himself undertook the task of clearly stating the creation of the Universe, its purpose, its destiny and the meaning and purpose of life. The Bible is the source of this knowledge.

We can spend a lifetime and more searching for the answers to the fundamental questions about the universe, but if and when we find these answers, they will be no more correct than those divinely provided. The acknowledgment of God is the key to all knowledge. The mere denial of God is enough to prevent man from gaining a true knowledge of the Universe, which is his natural right. So the scientific journey has been long and hazardous. We have come a long way since the days of Aristotle when the Geocentric Theory that the Earth was the center of the Universe prevailed. This theory prevailed for some time, even to Sixteenth Century. Since that time man has made gigantic leaps in the field of science. The theory that the planets orbit around the Sun was first proposed by the Polish priest, Nicholas Copernicus in the Fifteenth Century but was not taken seriously by the scientific community until sometime in the Sixteenth Century. The idea here is that the scientist needs the Theologian to help him address the fundamental questions which so far he has not yet successfully resolved.

There are limitations to science. Science cannot tell us all we need to know about the Universe, or at least has not adequately addressed the great cosmological question of space neither the origin of matter. Therefore we cannot expect it to address other questions having to do with the purpose and destiny of the Universe, the meaning and purpose of life, and the moral and spiritual code of human nature. Still there are a number of other important questions not addressed by science. Since it is the Theologian who has answered the fundamental questions about the Universe and life, it is hoped that he can help with the answers to other important questions about the Universe. The Theologian recognizes a God who from His infinite wisdom, love and power evolved the Universe with all its laws. Yet the Universe is not mystical. With the fundamental knowledge given by the Theologian we can gain more knowledge. This will necessitate the united effort of scientist and Theologian.

While it is obvious and logical that Science and Theology are naturally related and that this natural relation is inseparable, we have tried to separate them. And we can even think we have succeeded. In reality, they cannot be separated the one from the other. What we have done in practice is that, on the one hand, the scientist has studied nature without the recognition of God. On the other hand, the Theologian has studied about God with or without recognition of the scientific information of the scientist. This causes the scientist and the Theologian to oppose each other for most of the time. The end result of this is very tragic because the scientist will ignore the great truths of the Theologian and the Theologian will ignore the great facts about nature, its functions and its laws.

Logic dictates that the need for reconciliation between science and theology is most urgent and the purpose of having a better understanding of the Universe is more important than the cause of the difference between the two. It is against this background Theo-Cosmos must be seen. Our goal is to interpret scientific questions from a theological perspective to give a true and complete scientific description of the Universe. Modern science has provided us a long chain of information. Each link represents a solution to a mystery of the Universe. Still, modern science has failed to provide us a true and complete description of the Universe. As a matter of fact, with its evolutionary aspect assumed since the Nineteenth Century, it is further away from that possibility.

Indeed, the long chain of scientific information has not answered some of the other questions about the Universe. In some instances, some of the links of this chain have given rise to other questions that are equally important. We know that outer space is in a state of weightlessness. What is the cause of this? The answer to questions like this and others of like nature will aid us in our quest for a fuller knowledge of the Universe. But is the Theologian challenged by the scientific information of the incredible velocities of bodies in space, their distances apart, their sizes, the explosion of stars, the birth of new stars, the Big Bang Theory, the Theory of the Expanding Universe, the discovery of billions of galaxies like the Milky Way? Yes, the Theologian is challenged but not defeated because modern science only serves to illustrate his concept of God's infinity. And should there be a Theological Response to modern science? The answer is YES! The World has long awaited such Response.

Byran C. Russell

THEO-COSMOS:

A
Scientific Description

Of The

Universe

From A

Theological Perspective
Secrets Of The Universe
The Quest For The Knowledge
Of The Universe

CHAPTER 1
The Underlying Theological Postulate

A correct underlying Theological Postulate is imperative in any scientific description of the Universe from a Theological Perspective. Let us first define, THEO-COSMOS. The syllable "THEO" is derived from the word, THEOLOGY. Theology means the study of God and His relation to the Universe. "COSMOS" means the Universe. COSMOLOGY is the special branch of science that deals with the arrangement of the Universe. "THEO" is herein intertwined with COSMOS to convey a single intention —— a reconciliation between THEOLOGY and SCIENCE.

One of the Theological Arguments for the existence of God is the argument from CONGRUITY. The substance of this argument is that a theory which explains all the related facts in a case must be correct. The belief in the existence of an Omnipotent and Eternal Being explains the facts of man's mental, moral and spiritual nature as well as all the facts of the material phenomena of our Universe. "Belief" just cannot explain how a Being could be eternal and omnipotent. But "belief" is not contrary to reason, for we know that since there is a God, He must be eternal and omnipotent.

Science, today, is not what it used to be hundreds of years ago. Hundreds of years ago, science had a mild and gentle nature. Theology was then regarded as the QUEEN of science. Today science has assumed a new dimension – accidental evolution of the Universe and of life. Science has now documented the idea of accidental evolution by articulating several such theories. Thus science has broadened its role of studying nature and its laws, and has broken its natural band with Theology. The Theologian now feels the evolutionary aspect of science must be condemned by all. If Sir Isaac Newton and Albert Einstein, possibly the two greatest scientists the world had ever known, were alive today, they would have condemned the evolutionary aspect of modern science. Sir Isaac Newton is reported to have said, "Space is an attribute of God". Albert Einstein is reported to have said, "God does not play dice".

In the past, Theology emphasized God's relation to the Universe while neglecting scientific details, even those in the Bible. This neglect has had far reaching negative consequences; and to a great extent, has paved the way for Accidental Evolution Theory of the Universe. It is true that Creationism has been preached since the beginning of time. But what references have been made to the field of science and to those scientific references in the Bible? How relevant are the scientific findings and information to Theology and to the field of science?

If one has no opposition and is unchallenged, one is more likely to do as one pleases. Scientists will tell you that three centuries ago, philosophers would question scientific reports. This caused the scientists to be more cautious in their reports. But while the philosophers were doing that, what was the Theologian doing? They were saying absolutely nothing. There is much information in the Bible from which the Theologians could have formulated a theory of the formation of the Universe. Thus the Theologians have had more to their advantage than the scientists. The Author feels science is very important to the field of Theology. The evolutionary aspect, the great cause of the disharmony can be examined and made to reconcile with

Theology. The hope of reconciliation lies in the fact that accidental evolution teaches that the Universe is not eternal. It shows in a scientific way how something came out of nothing. The evolution of life theory shows that the origin of life is a mystery because it has failed to satisfactorily explain how life began.

It is unfortunate that scientists have formulated these theories. We do not condemn the scientists. We condemn the accidental aspects of the theories. Our aim in THEO-COSMOS is to present a World View, a picture of the Universe which will give one a better understanding. With this aim, scientific theories and information will be examined. Our underlying Theological Postulate should not disappoint certain expectations of science. At least, science should expect that this Postulate is in keeping with its concept of an infinite Universe, or this Postulate should accommodate an infinite Universe. On the other hand, this Postulate should not disappoint certain expectations of the religious world. One suspects that to do the latter will not be without much difficulty, since there are so many religions with great diversities of beliefs. A true underlying Theological Postulate must be based on a true interpretation of the Bible. It must address the Bible itself, God Himself, and the works of God— CREATION.

The Bible

The Bible is the text book of the Christian Religion. It is undoubtedly the best known book, the best read, and the oldest text book. Its universality is accompanied by an inescapable justification. There must be a reason for its universality. The reason is its Divine Authorship. A book expresses the mind, the thoughts of its author. The Bible in its essential nature expresses the mind of God regarding His purpose in Creation, the Fall, Redemption, and the Destiny of Mankind. And how could we know the mind of God except by Divine revelations.

The predictive aspect of the Bible also supports its claim of Divine Authorship. Many predictions of future events were made, most of which have already been fulfilled. Some are yet to be fulfilled. The prophets who made these predictions did not claim they were the results of their natural insight but were due to instantaneous endowment of supernatural knowledge. The claim of Divine Authorship does not mean that most of the information of the Bible is comprised of revelations or predictions. It means the most fundamental information was derived by Divine revelations. Man could not have known about the attributes of God except by Divine revelation, neither about Creation, the Fall and the Destiny of mankind.

The Bible for the most part is a record of God's relationship with mankind. In light of this, the Bible is naturally cumbered with secular details, man's behavior and God's response to that behavior. Much of the Bible expresses man's despair as well as his courage, faith and determination to survive. Much of the Bible clearly mirrors man's true nature so that one can predict how man, under given circumstances, will act and react But the claim of Divine Authorship is also supported by the indestructibility of the Bible, its character and its unity. Though the Old Testament Scriptures were officially accepted as the revealed and inspired Word of God by the Church Council A.D. 200 and the New Testament Scriptures A.D. 400, the Bible dated back to a much earlier period to God's first dealings with man. We know that God's dealings with Abraham dated back to at least 4000 years. Only 11 chapters of the entire Bible predated God's dealings with Abraham. That period would have further added to the age of the Bible. But how many books are that old? If there is one, does it have any relevance to contemporary man as the Bible does?

The Bible has not survived under the best of circumstances. There were always enemies of the Bible, people who hated the moral and spiritual demand it required of them. There were people who hated its strong condemnation of sin and evil. And these were always people in seats of authority. The Bible has survived relentless and merciless attacks aimed at its destruction. History records for us the unleashing of the wrath of the Roman Empire against the Bible. This was in no limited measure. In A.D. 303 Diocletian the Roman Emperor made a decree that every copy of the Bible be destroyed by fire. All the available Bibles were burned. But this effort did not eliminate the Bible. Neither did it discourage those who believed in it. Diocletian died and his decree also died. Constantine who succeeded him, after a time, never felt obligated to this decree. As a matter of fact he had a different attitude.

Constantine's attitude was influenced by a Divine experience. He was going to war when he looked to the sky and saw a cross, the symbol of the Christian Religion. Seeing the cross, he heard a voice saying, "By this sign you shall conquer". He knew it was the voice of the God of the Christian Religion. He responded to that voice and was converted. One memorable result of his conversion was that he made the Christian Religion the state religion.

Today there are many Diocletians, people in all walks of life who hate the Bible. Today there are even some governments that will not permit the use of the Bible in their countries. We can easily know what countries by their comparative backwardness. The Bible survives with astounding relevance and influence at the heart of modern civilization.

While the secular details are everywhere in the Bible, the emphases are placed on such subjects as God's Nature and Attributes and Sovereignty, the Fall of Man, the Consequences of Sin, the Immortality of the Soul, the Power of Man's Freedom of Choice, the Plan of Redemption, Man's Destiny, the Second Coming of Christ, the Final Judgment, and the Eternal State. Even the secular details are not without purpose. Besides revealing our true nature and character, they are the narration of an intricate love story between God and mankind. Only as we see these secular details in this light can we understand why a Holy God has preserved the human race.

The knowledge of the secular details is of much practical value to us. So in the truest sense, the Bible is the very fount of knowledge. We carefully note that the Bible records creation of the Universe and man, the first marriage, the first responsibility of man and his failure in that responsibility. The Bible records the first birth, the first death, the first encounter of man with God after he had sinned and the immediate results, shame, expulsion from Eden, labor and sorrow. The Bible records such Divine judgments as the Destruction of the Antediluvian World, the Confusion at the Tower of Babel and the Destruction of Sodom and Gomorrah by fire. Finally the Bible records the Exodus of Israel from Egypt. Today the existence of the Jewish Nation in Palestine is strong evidence to the Divine character of the Bible.

The Divine Authorship of the Bible can be also perceived from its majestic unity. The compilation of its 66 books took approximately 1600 years. These books are said to be written by 40 different writers. In one sense the first five books and also the sixth were directly written by God because of the personal and direct verbal communications between God and Moses and between God and Joshua. This can also be said of the book of Revelation. The other books can be said to have been indirectly written by God, though most of those books contain Divine revelations and the writers were divinely inspired. When we look at these books individually and ask the question, "What are their underlying messages?", we arrive at the same answer as when we ask the question, "What is the underlying message of the Bible?" This excellence of unity far transcends that of any other book.

We cannot ignore the fact that some details in the Bible can be misunderstood. These details are more easily misunderstood when we read the Bible with preconceived notions instead of an open mind and a receptive heart. But we can see them as a motive for searching the Scriptures constantly. John Wesley was right when he said the Bible could not have been written by bad men or bad angels, good men or good angels. He explains that the Bible condemns the souls of bad men and angels to hell. So why would they write the Bible? Good men and angels could not write the Bible because they would not have said,

"Thus says the Lord" when it was they who said it. Therefore the Bible could only be written by God Himself. We must here state that the Theological Postulate with respect to the Bible means that the information of God, His nature, attributes and works contained therein is absolutely true. We shall now look at the Theological Postulate with respect to God.

With Respect To God

A reasonable understanding of our Universe hinges on the realization that God is the indispensable factor in all of knowledge. Yet the diverse views and ideas of God would naturally belie the truth of this statement. Here one would be forced upon a most precarious path fraught with great uncertainties. And the religions do not help to alleviate the pains inflicted on the journey of this path, because religion must first answer the question so frequently asked among them, which of us is the right religion? While the major religions share the burden of declaring the underlying message of the LIFE BEYOND, their ideas of God make them fundamentally different. Even in the Christian Religion all are not agreed on the doctrine of the Trinity. From our point of view, it is important that we here and now articulate the nature and attributes of God. This is critical to a true scientific description of the Universe. The only authentic source of this knowledge is the Bible. The Bible clearly teaches:

1 The Eternity and Self Existence of God. The Bible does not argue or attempt to prove the existence of God. It has assumed the existence of God and that mankind believes there is a God and that this belief is innate in man. You see, when the Bible was written there were no doctrines of evolutions. By the eternity of God is meant that He is without beginning, that there was never a time when God was not existing. He always was, is and shall be. We do not pretend to understand the eternity of God beyond this level. We are satisfied that God exists and because

He exists, He must be eternal. By the self-existence of God is also meant that His existence and nature are independent of matter as we know it. As Eternal, God transcends time.

2. The Infinity of God. By the infinity of God is meant that He is free from all limitations. In the first place, He is not bound to time. In the second place, He is not bound to space. He transcends space and time. In the third place, He is not limited in any of His attributes (spiritual infinity). An Infinite God can limit His manifestation. This quite often is the case. Infinity in His attributes means that He is omnipotent (all powerful), omnipresent (everywhere present), omniscient (all knowledge and wisdom); and His love, justice, mercy, holiness are without limits.

3 The Spiritual Nature of God. By the spiritual nature of God is meant that He is a spiritual substance. He is not made of matter as we know it. Due to His spiritual nature He is immortal, immutable and invisible. Being immortal He can never cease to exist. Being immutable His attributes cannot be changed or improved one way or another. Being invisible He cannot be seen by the human eye. He can, however, manifest Himself in a form that the human eye can recognize. The eternity and self-existence of God, His infinity, his spiritual nature— all imply that He is the LIVING GOD, that He had the ability to create the Universe and He did.

4 The Character of God. By the character of God is meant the sum total of all His Divine activities and actions. All Divine activities and actions must be rooted in His holy nature. Herein lies the foundation of the moral Universe. The Moral Universe implies accountability. All moral beings must be accountable to the Creator. The Creator cannot do anything that is wrong.

Because man was created in the image of God, he is always inclined to do that which is right, even though he often does that which is wrong.

In our Moral Universe the actions of men are always questioned as to whether they are wrong or right. Their actions of wrong or right determine their character. To us, therefore, character is of utmost importance. The Bible contains a number of Divine actions which seem to have contradicted our definition of the character of God. While we cannot enumerate all of them, we wish to mention some of the most obvious and fundamental ones in relation to mankind. We here mention the following ones:

1. SIN. God is not the Creator or the Cause of sin. Sin resulted from the Free Choice of man in rebellion to a known law or a Divine command. Man's Free Choice is the faculty of his soul that elevates him above the level of the animal creation and makes him like his Creator. This Free Choice with which man was created must be considered a holy act on the part of God. Today sin is universal because we sin by nature and by choice.

2. The Destruction of the Antediluvian World. The physical destruction of mankind except for 8 persons was a judicial act of God, simply because mankind had consistently revolted from God in sinning beyond the TOLERABLE LIMIT. God has a sovereign right to rule the moral Universe but can only do so in keeping with His own moral nature. His moral nature dictated the destruction of the Antediluvian World and yet its preservation by saving 8 persons. The surprise is not really about the destruction of the Antediluvian World. It is really the tolerance of God's holy nature of sin to the degree that He permitted them to sin for years.

One may form the same conclusion of some other Divine judicial acts such as the destruction of Sodom and Gomorrah, the destruction of the Egyptians in the Red Sea, and the wars of Israel with the nations of Canaan. All these must be seen as Divine acts of a punitive nature executed in order to preserve the human family, in keeping with the Divine promise of Redemption. We must understand that the holy nature of God cannot tolerate sin beyond a certain limit. One can also see that the Divine choice of the nation of Israel was for the purpose of a specific role in the plan of Redemption and was not an exercise of mere preference of one nation above the others.

Our conviction remains rooted in the truth that God's character is the foundation of our moral Universe and that foundation is firm and steadfast. Today men act and react in view of moral ends. And if such actions and reactions are wrong, by their faculty of conscience, they feel self-condemned. Indeed, a sense of guilty conscience causes them great human pain. The knowledge of the self-revealed personal God of the Bible is indispensable to a complete human *understanding* of the physical Universe. We now turn our attention to the Theological Postulate with respect to the works of God.

The Works of God – Creation

The Theological Postulate is the heart and soul of the **Theo-Cosmos** because without it, reconciliation between science and Theology with the aim of giving a true scientific description of the Universe is not possible. And the heart of this Postulate is Creation – the works of God. The Universe is an effect; every effect has a cause. An intelligent cause will produce intelligent effects; accidental cause can only produce accidental results.

Since all scientists admit the intelligent and systematic arrangement of the Universe, its cause must be intelligent. God must, therefore, be the cause. The original definition of science, in its simplest form, is the study of nature. Like many things in our World today, science has lost its simple, pure, and original definition. With the discovery by Edwin

Hubble of the motions of galaxies at the outer reaches of the Universe in the 1,920's, science has assumed an additional meaning —— the meaning of accidental evolution. It is generally characterized by this new meaning.

Some scientists estimate the observable Universe to be 200 billion to 500 billion galaxies, each having 100 billion stars. Evolution explains that all of this evolved accidentally. Theology, on the other hand, assumes the whole observable Universe and the various life forms on Earth to be the direct creation of God. However, Theology has failed to propose a theory as to how God might have formed the Universe in spite of the known laws of science, the many scientific references in the Bible, and the knowledge of God's nature and attributes. Since man was created in the likeness of God, he can be expected to propose a theory as to how God might have formed the Universe.

Let us look at it this way——God has revealed Himself and His inner nature to man because he has the capacity to perceive Him. However, without the self-revelation of God, man could not perceive Him thus. Nevertheless, man also has the ability to focus his attention primarily on the material and secular things of life instead of searching after God who is comprehensible. Even though He has revealed Himself so glaringly, the other animals cannot perceive Him the way man does. The point is that man has the intellectual endowments to perceive God. With this intellectual endowment he can discover the laws of nature with or without Divine revelation. God created the Universe for man to understand about it. Some Theologians think that the whole visible Universe instantly came into being in its present form. In **Theo-Cosmos** we are taking a different position. We recognize two great contradistinctions in the formation of the Universe– the Miraculous Aspect and the Scientific Aspect.

The Miraculous Aspect

This aspect can be identified but not explained. The Miraculous Aspect was the creation of space and the conditions of matter: heat and cold. These conditions imply the idea that matter was in its most elementary form. After the creation of space, God spoke these conditions into existence. The whole special Universe was instantly subjected to intense heat: matter in the form of hot gas. The creative power of God to effect these conditions cannot be explained. And in **Theo-Cosmos** we are more interested in the effect of that power, Creation, than the power itself.

The Miraculous Aspect of Creation is completely ignored by the evolutionary aspect of science. Yet the formation of the Universe as explained by Evolution is in itself miraculous. Herein lies the possibility of **reconciliation**. Science (evolution) and Theology cannot both answer the fundamental questions of the Universe. Indeed, science (evolution) has failed to answer the first of three most fundamental questions. It has not answered the question of the cause of the Universe for the simple reason that an accidental cause cannot produce an intelligent effect as the observable Universe. In the second place, it has not answered the question as to the purpose of the Universe, why is it here? In the third place, Evolution has failed to answer the question of the moral and spiritual nature of man. Theology has adequately answered all these questions. Thus, science must be reconciled to Theology.

The Scientific Aspect

At a certain point, the Miraculous Aspect assumed the laws of science. Following the creation of the conditions of matter, one could follow the stages in the formation of the Universe according to the laws of science. At this point, the premise can be established that every phenomenon in nature, event, occurrence or condition is the result of some other phenomenon, event, occurrence, or condition in nature. However, it would not be prudent to predict a time table of all the stages in the formation of the Universe. It would be a fatal mistake to set God a time table. It is satisfactory to say the formation of the Universe was not instantaneous: it involved an indefinite period of time. This is not placing limitation on God's omnipotence; it is rather articulating a choice in the exercise of that omnipotence.

In their book, Science Matters, Hazen and Trefil emphasize that creationists claim the Earth is 10,000 years old. I quote, "Biblical Creationists accept on faith the literal Old Testament account of Creation. Their belief includes (1) a young earth, perhaps less than 10,000 years old; (2) catastrophes, especially a world-wide flood, as the origin of the earth's present form, including mountains, canyons oceans and continents and (3) miraculous creation of all living things, including humans, in essentially their modern forms" Science Matters, Doubleday, 1991 Ed., p243 I personally had not known that religion or creationists made such claim or expressed such belief. If such claim made or belief expressed by religion, it is with Biblical foundation, or an incorrect interpretation of the Bible. The book of Genesis begins with an absolute statement of truth: "In the beginning God created the heaven and the earth". But no one seems to know "The Beginning". The statement of verse 1 seems to have referred to Creation in the Dateless Past of which no details were given .and seems to suggest that during that time the Earth remained uninhabitable. God had a lot to do to make it habitable as indicated from verse 2 to verse 31. No Theologian knows how long ago that was or how long a period was involved in the creation of the Universe in the Dateless Past. The scientific age of 15–20 billion years is the time the evolutionists say the Universe took to evolve.

Verse 3 of Genesis chapter 1 is not directly related to the opening statement of absolute truth in verse 1. Verse 2 begins to tell us of specific developments in the Earth, that in 6 days God created the life conditions and life forms of the Earth. Notice that when He began the creation of life conditions the Earth was already covered with water. The Earth was already developed except for these conditions. Religion has never attempted to explain how long the Earth remained uninhabitable after it was created in the initial act of Creation in the Dateless Past.

Because the Bible did not say how long it took God to form the Earth and the Universe, the theologian is not refuting the scientific age of the Universe. The date of human civilization is another matter. No one knows how long ago Adam and Eve our first parents were created nor at what point the record of human civilization began to be kept. The age of the Earth and the age of human civilization are two different subjects and must remain that way.

The authors of Science Matters think that there are two strongly held views about the Universe and life: Evolution and Creationism. But how strong is strong, how strong is the view of evolution compared with creationism? In all honesty, evolution is a very weakly held view. There are many scientists who believe there is a God. As recent as January 3, 1991 the St. Petersburg Times of Florida carried the headline "Big Bang Theory Turning Out To Be Big Bust". It states "A critical element of the widely accepted Big Bang Theory about the origin of the Universe is being discarded by some of its staunchest advocates, throwing the field of Cosmology into turmoil." The following day a leading scientist was reported by television news as saying, "The Universe was manufactured, meaning it was created."

The major religions of the World: Hinduism, Islam, Judaism, and Christianity, with millions of followers each, reveal the strength of Creationism. Besides, there are millions of people who have no expressed any particular allegiance to one religion or another who believe there is a God. It is rather frightening that a few scientists (evolutionists) do not want Creationism to be taught in the public schools of America. If evolution is so strongly held a view, and if they are so confident about it, why not allow it to be tested in these institutions against Creationism? It would be a good thing if Creationism was taught in all public schools throughout the World and in particular, in the United States of America. But I am not really bothered about this strategy of the evolutionists because it will have little or no effect on the minds of the children. Before the children begin to go to school, they know there is a God. And when they graduate they will tell the professors and the World that evolution does not make sense.

The Scientific Aspect of Creation implies that all the laws of science satisfy the desire, will, and purpose of the Creator. This statement is profound. Only as we understand it, can we understand God's purpose in Creation. The Creative Spoken Word of God systematically held the Universe together. This does not minimize the fact of the immanence of God, that He over-sees nature. A human scientific invention is a good analogy. A scientific invention was invented according to those scientific principles peculiar to that invention; and an invention functions according to those principles. The inventor still oversees his invention with or without intervening in its functioning. He can intervene at any time. If he does, he has good reason. Not intervening does not mean he has abandoned it. It only means an intervention was not necessary. This is the same kind of relationship God sustains with the Universe.

One of the great things about nature is its variety. And underlying this variety is the kind of immutability that makes us curious. Why does matter retain three distinct forms (solid, liquid, and gas)? Why the motions of the Solar System are so consistent? Why the Earth produces different kinds of plants? Why the Earth produces millions of different kinds of animals? The answers to these questions can be only justifiably explained in the knowledge that the laws of nature work to satisfy the desire, will, and purpose of the Creator.

Creation is the sublime revelation of God to man. The intelligent construction of the Universe reveals His omniscience. The immutability of the laws of nature reveals His immutable nature. The moral and spiritual nature of man reveals His moral and spiritual nature. Still, man a finite being cannot perfectly know everything about God, an infinite Being. The same truth applies to man and his quest to know about the Universe. He can have a general understanding but he cannot perfectly know all. Yet that which he can know must be within the limits of a Theological Context. Man cannot find out the cause of the Universe by scientific means – that is certain. He knows the chemical properties of life but cannot put them together in a manner to produce life. Without the recognition of God, man knows very little about his own nature and in particular not much about the Universe.

Creation is the most important Divine act. Some may think Redemption is the most important Divine act. But if God did not create the Universe there would be no need for Redemption. The Creator wants us to know all we can about the Universe. He could have evolved it without speaking a single word. He did not because He intended for His creatures to understand. He created man with an infinite capacity for knowledge. Besides, our fore-parents Adam and Eve chose knowledge instead of life. What we are today seeing in science and technology is the result of that KNOWLEDGE. It was the conjunction of that knowledge — **disobedience** was wrong.

Because of the intellectual endowment of man, he would in time deduce from nature a theory as to how the Universe was formed by God. From a knowledge of the nature of God, the laws of the Universe and from the scientific references in the Bible such a theory can be formulated. Such theory is not guarantee that an Omnipotent God had to form the Universe that way. It is only a possibility. And to deny a theory that possibility is in itself a denial of Omnipotence. In the next chapter the **Theo-Cosmos Theory** will be proposed.

All Christians believe that God created the Universe, but many ignore the laws of the Universe. In creating the Universe, God created its laws. Therefore, to understand Creation, we must also seek to understand its laws.

CHAPTER 2
Scientific Theories: Review and Response

We can learn a lot from scientific theories. And true learning is like a chain reaction: it stimulates more knowledge. In the process of man's quest for scientific knowledge, he has formulated a number of scientific theories. Some can be considered the epitome of human folly. Some can be considered a triumph of human intellect.

Reflecting on scientific theories has great advantage to him who desires knowledge. One can thus determine the strength and weakness of a particular theory. Another good thing about scientific theories is that they summarize a wide range of knowledge. This is good for everyone. The scientific theories we are about to review can do no less and can be of no less significance to us. They will deal with matter in its two extremes: The MICROCOSMIC and the Natural Phenomena. We shall begin with the more well-known theories of the origin of the Universe.

The Hot Big Bang Theory

In 1927 American astronomer Edwin Hubble discovered that

galaxies at the outer reaches of the Universe were moving away from each other at incredible speeds. This discovery had three major effects on the scientific community. One of these effects was the abandonment of the **Cosmological Constant**, a theory proposed by scientist, Albert Einstein during 1915. The second effect of Hubble's discovery was the formation of the theory that the Universe was expanding. The third effect led to the formulation of one version of the Hot Big Bang Theory. Scientists reasoned that if the galaxies were moving farther away from each other, then at one time they must have clumped together and the Universe was one gigantic mass. They reasoned that there must have been a cataclysmic explosion at the beginning of the Universe whose tremendous force was still actively pushing the galaxies farther apart. This is the background to one version of the Hot Big Bang Theory.

The explosion of the Cosmic Atom 100 million miles in diameter marked the birth of the Universe. About 20 billion years ago this super-large atom of protons, neutrons, and electrons exploded scattering its mass into space. Minutes after the explosion, temperature exceeded several billion degrees. Particles began to form nuclei. About 30 million years latter temperature had fallen to a few thousand degrees and some of the gas contracted into dust. The force of Gravity began to attract the dust into great masses. From these masses the galaxies of stars originated. This theory is attributed to French Scientist, Lemaitre (Julien de La Mettrie).

The Inflationary Version of the Hot Big Bang Theory

The theory states that within a fraction of a second after the Big Bang, the size of the Universe increased billions of times. The Universe then settled down to its present rate of expansion. This theory was first proposed by Dr. Allan Guth, a scientist at the Massachusetts Institute of Technology.

The Hot Big Bang Theory As Stated In A Scientific Report

carried by the St. Petersburg Times (Florida, U.S.A.) April 24, '92 Under the **Head-line**: RELICS OF THE BIG BANG FOUND.

In 1989 NASA launched into orbit a satellite called the Cosmic Observer Background Explorer (COBE). Its mission was to detect clues of the origin of the Universe. The report reads: "Scientists say evidence found at the edge of the Universe explains how galaxies were formed and supports the Big Bang Theory of the creation of the Universe." The report continues. "Ripples discovered at the edge of the Universe have provided the first evidence of how stars and galaxies were formed, scientists say."

This scientific report could only be regarded as the most misleading scientific report in the history of science because these relics were merely variations in the Microwave Radiation Background (the temperature of outer space of 454 degrees below zero). It was praised by some of the leading scientists involved in the project. "If you are religious, it is like looking at God", said George Smoot an astrophysicist who headed the research and the spokes-man for the report. "What we have found is evidence for the birth of the Universe and its evolution," he continues.

Joel Primack a physicist at the University of California, Santa Cruz said: "It is one of the major discoveries of the Century. In fact it is one of major discoveries of science." Michael Turner a physicist at the University of Chicago said: "The discovery is unbelievably important. The significance cannot be over stated. They have found the holy grail of Cosmology.

If is indeed correct, this certainly would have to be considered for the Nobel Prize." The report continues: "What scientists discovered were large regions in the Universe in which the temperature varied by no more than a hundredth thousandth of a degree from the areas around them. "George Smoot explains the variations in the Microwave Radiation Temperature (the temperature of 454 degrees below zero of outer space) thus: "The temperature differences represent different densities of matter in wispy clouds and surrounding regions. These small variations are the imprints of tiny ripples in the fabric of space time put there by the Primeval Explosion Process. Over billions of years, the smallest of these ripples have grown into galaxies, clusters of galaxies, and the great voids of space."

Before we respond, let me restate the Big Bang Theory carried by the Report: The Big Bang occurred about 15 billion years ago. The entire Universe explodes out of a point of infinite density. The explosion is uniform and featureless.

Inflationary Expansion. The Universe expands dramatically. Matter in the Universe is still evenly distributed. 300,000 YEARS AFTER THE BIG BANG. Huge clouds of matter, 500 million light years in length and larger beginning to condense.

Galaxy Formation. The first galaxies appear about 200 million years after the Big Bang.

TODAY. Fifteen billion years after the Big Bang, stars and galaxies have evolved out of the clouds of matter. The lights from events that occurred near the start of the Universe have traveled for about 15 billion years to reach the Earth.

Response. How do we respond to the scientific discovery of the Cosmic Background Explorer Satellite? What can we say? First of all, any effort of man to gain a better understanding of our Universe must be commended. In particular, the scientific research of COBE is indeed courageous and equally ambitious. The government of the United States of America and the scientists must be congratulated for even the mere attempt. However, it must be emphasized that we are dealing with reality and we are dealing with it at its heart. The heart of reality is that, from a human point of view, we live in an infinite Universe, in which there are all kinds of variables, so much so, that things many times appear to be what they are not. First, the light from those distant stars would not necessarily reach the Earth because light travels in a straight line. Therefore, only the light of stars aligned with the Earth would have reached. For a good while mankind thought that the Earth was the center of the Universe. Our Cosmology of today is a radical departure from the old Geocentric Theory. The concept of the Geocentric Theory was due to the fact that things genuinely appear to be what they are not. And the radical departure from that theory is due to the fact that things are what they are, and not what they appear to be.

The discovery of temperature variations in the Microwaves Radiation Temperature (the temperature of outer space 454 degrees below zero) is not a certain clue as to the origin of the Universe, but is a natural occurrence in the infinity of space. As a matter of fact "variations" are a natural law of nature and mothered in the size and structure of matter. On Earth, locations within very close proximity have varying temperatures and only areas within a particular location have the same temperature.

Even if the Microwaves Radiation was the glow resulting from the Big Bang, at some point in the infinity of space that glow or radiation would naturally vary in temperature. While we are not assured there was any Big Bang, for that matter, we are absolutely sure the innumerable stars convert countless tons of matter into radiant energy. The Universal Law of the Conservation of Energy, that energy cannot be created nor destroyed, suggests that the Microwaves Radiation must be an effect of combined star radiation (cosmic radiation). In a subsequent chapter (The Economy of Cosmic Radiation) more will be said. It was here expedient to state that variation in Microwaves Temperature is not a phenomenal discovery.

This most recent version of the Hot Big Bang Theory has left room for wild speculations as to the origin of space. It has not given a definitive statement about space. Space would have to be a prerequisite of that Primeval Atom of 100 million miles in diameter. The second flaw of the theory is that it has said nothing as to how the matter of this Primeval Atom originated. For these two fundamental reasons there is not much substance to the Theory. And by virtue of these inherent deficiencies, the Theory has not given us the cosmological secrets of the Universe.

As we have quickly pointed out the flaws of the Theory and without hesitation rejected them, we must concede the favorable features. We see in the favorable part of the Theory the equivalent of a miracle. For matter that volume and infinite density to have exploded could only be a miracle. The scientists have unconsciously admitted this. We know that the splitting of the atom by man took diligent and systematic efforts over a long period of time. The explosion of the Primeval Atom implies external intervention. This atom when compared to the size of the Universe could only be regarded as less than nothing. The development in the evolution of the Universe could also be regarded as a miracle. So in essence, the Theory shows how something was made out of nothing.

We take this position of reconciliation because to utterly reject the Theory would be placing severe limitation on an Infinite and Omnipotent Creator. Yet we cannot say with any degree of certainty there was Big Bang. To say that God could not create a mass of matter of infinite density and then exploded it in forming the Universe is, indeed, the placing of limitation on the Omnipotent Creator. But I am fully convinced that by correctly interpreting the scientific references in the Bible of Creation and of the nature of God in conjunction with the correct interpretation of the laws of science, we can perceive exactly how God formed the Universe.

If one believes the whole visible Universe came into being instantly in its present form, then there is nothing else one can know about its formation. Many Christians including Theologians share this belief. Dr. Thiessen author of Introductory Lectures in Systematic Theology shares this belief. And I suspect many other outstanding Theologians share this belief. I now quote Dr. Thiessen: "By immediate creation we mean that free act of the Triune God whereby in the beginning and for His own glory, without the use of preexisting materials or secondary causes, He brought into being, immediately and instantaneously, the whole visible and invisible Universe". <u>Lectures in Systematic Theology, Wm. B. Eerdmans Publishing Co., 1963 Ed., p 161.</u>

Dr. Thiessen is regarded as one of the greatest Theologians of all times, but I beg to differ with him on the subject of Cosmology. The scientific references in the Bible all suggest that the creation of the Universe was not instantaneous. The Genesis Account of the creation of life forms and life conditions of Earth suggests the progressive nature of Creation. When one reads the 27th verse of chapter 1, one gets the impression that Adam and Eve were simultaneously created. Then when reading chapter 2, one discovers that Eve was created sometime after and exactly how she was created.

In Job 38:4-7 we read: "Where was thou when I laid the foundation of the earth…When the morning stars sang together and all the sons (angels) of God shouted for joy?" This reference clearly shows that some parts of the Universe are older than other parts. This Scripture reference brings to mind a father who takes his sons to watch him play his favorite game or to do his favorite exercise. Here we see God taking a number of His angels to watch an awesome display of His omnipotence. Having watched it, the angels shouted for joy. There are other Bible references that suggest the progressive nature of the formation of the Universe. I cannot see why God an eternal Being would be in a great haste to create an instant Universe knowing full-well that shortly after it would be marred by sin. Anyone who believes in instantaneous Creation should think again. Even the creation of man was a process. God formed man from the dust of the Earth, and breathed into his nostril the breath of life.

Having exposed the flaws in the Hot Big Bang Theory and having explained the progressive nature in the formation of the Universe, we shall now propose a theory that will give the true Cosmological picture and unlock to mankind the secrets of the Universe. Such a theory must address the two great CONTRADISTICTIONS in the formation of the Universe: the Miraculous Aspect and the Scientific Aspect. It must give a definitive statement on the origin of space, the origin of matter, the weightlessness of outer space, and the formation of stars and their motions. The intelligent arrangement of the Universe suggests it was carefully preplanned and followed a process of development.

The Theo-Cosmos Theory of the Formation of the Universe

The Miraculous Aspect. This aspect in the formation of the Universe had to do with the creation of the conditions of matter: space, heat, and cold. Empty space was first to be created. Before the Creation of the Universe, The Eternal God existed in the fullness of all His attributes. One of His attributes was an infinity of majestic Light. In consideration of Creation, He limited the manifestation of that majesty. The immediate result was empty space. An Infinite Being can limit His manifestation.

The picture that comes to mind from understanding that matter exists in three forms (solid, liquid, and gas) is an interplay of heat and cold, expansion and contraction, and Gravity and Motion within the background of weightlessness. The whole cosmological secret is SPACE. If there were no space, there could be no infinite density in the first place. If there were no space there could be no conditions of heat and cold.

Knowing about space as we now know, it is almost impossible to think that there was a time when there was no space. Is it not? But there was a time when there was no space. Now close your eyes and close the Universe out of your mind and think you were the only person. As you remain with your eyes closed and think that nothing else exists, you are actually picturing the Cosmological situation before Creation. Before Creation there was only one Infinite and Eternal Being, the CREATOR who existed in all the fullness of His attributes.

The Scientific Aspect. Following the creation of space, God then commanded the other conditions of matter into existence. The whole spatial Universe was subjected to intense heat — matter in the form of hot gas. This statement is implied in the Book of Job 38: 37-38. From this point, the Universe began to form according to the laws of science. Most of the hot gas cooled and contracted into **cold invisible matter.** The rest cooled and contracted into clouds of dust above. The **cold invisible matter** far exceeded the clouds of dust. The clouds of dust then floated on. With the far greater amount of **cold invisible matter,** the condition of weightlessness was created — the foundation of the Universe was established.

Apparently it seemed unlikely that the greater amount of hot gas should have cooled into COLD INVISIBLE MATTER and the less amount into clouds of dust. But why when a cup of hot milk cools the cream is thicker than the rest of it? Less than one percent of that cup of milk formed into cream. We could think of the clouds of dust as the Cosmic Cream at the beginning of the Universe.

This formative stage of the Universe is clearly implied in Job 38:37, 38: "Who can number the clouds (stars) in wisdom? Or can stay the bottles of heaven, when the dust groweth into hardness and the clods cleave fast together?" Both the early formative stage of the Universe and the theory of star formation are here implied. If you had read the entire chapter you would have observed that the Almighty was giving Job a science lecture in which Astronomy and Cosmology were among the main points. Notice that in verses 31 and 32 the emphasis was on Astronomy. The Almighty lectured Job on four constellations of stars. "Canst thou bind the sweet influences of **Plades**, or loose the bands of ORION? Canst thou bring forth **Mazzaroth** in his season? Or canst thou guide **Arcturus** with his sons?" Then in verses 37 and 38 the Almighty was plainly saying to Job, "Did you know that all the stars of heaven were once in a state of dust, but at My command they were formed into stars of light?" God should not be limited to one method of Creation, and Instantaneous Creation seemed to do just that. human logics cannot determine God's method of Creation, but Progressive Creation does not seem to be contrary to logics.

Following the contraction of gas into clouds of dust, God separated huge volumes of matter into galaxies. He then initiated the formation process proper by first forming a super-star or stars at the center of a galaxy. The super-star or stars would begin a chain reaction in the galaxy and other stars would begin to form. If we picture at this point, the laws of science in effect, the galaxy of matter would begin to break up into smaller volumes at incredible speeds. As the process continued, greater density, temperature, and motion would be acquired.

The Principle of Center Formation initiated at the center of a galaxy would also apply to the formation of individual stars. The principle is to create the greatest force of Gravity within the shortest time. This would be absolutely necessary to speed the formation of stars. Without this principle of Center Formation stars and other bodies would not be spherically shaped.

During the formation of the first super-star or stars at the center of a galaxy, the rest of the galaxy would be contracting. Upon the instant formation of the first super-star or stars, immense gravitational force would be exerted upon the rest of matter in the galaxy. The instant effect would be that the unformed matter would rush towards the Center of Gravity, the super-dense star or stars. In the process, the unformed matter would have broken up into billions of smaller volumes and would have begun rotating.

Early in the process of a star's formation, it would begin burning at the center of the volume of matter from which it was formed. Matter would be most dense at the center and would thus begin to burn. This burning would not be destructive because it would attract surrounding matter of clouds of dust and gas at incredible speed. The surrounding matter would be overwhelming and while fueling the burning center, would control it by its great volume. This means that the volume of clouds of dust and gas would far out-weigh the center Burning. It is like burning a large amount of trash. If one starts a fire at the center for a long time, the surrounding trash would fuel the fire and at the same time systematically control it.

The process of burning and contraction and motion exerted tremendous pressure on the center. This would cause it to become super-dense and hot. When the whole volume of matter becomes a part of the formation process, a star is born — the star acquires full size temperature, density, velocity of motion and orbit. The inner core would remain super-dense and the outer parts would continue burning. If a star begins burning after its formation, it would have burnt disproportionately because of its size and would have thus quickly exploded. A star lives by burning and so burning has to be critical to the whole formation process. The orbit of a star would be within the circumference of the volume of matter from which it was formed. Rotations and orbits and their velocities were predetermined by the formation process.

Special Emphasis.

The **Theo-Cosmos Theory** recognizes that the formation of the Universe was not an instantaneous Divine act. This is not putting limitations on God's omnipotence or questioning God's ability to create an instant Universe. The Theory recognizes the formation of the Universe as a progressive Divine act and that parts are older than other parts. But the Theory will not set a time table based on any scientific calculation as the age of the Universe or the duration of its formation. Many things are older than the process of their development or completion. A house 20 years old might have taken less than a year to build. The formation of the Universe was a CREATION. It took the Creator less than a day to form man from the dust of the Earth and to breathe into his nostril the breath of life. It took the biological laws of nature 9 months to produce an infant and 20 years for that infant to become a man.

Conjecturing about the age of the Earth and the Universe does not add neither subtract from it. Assumptions cannot be proven. The answers of science to most of the fundamental questions about the Universe and life are based on assumptions. That is why science has more questions than answers. Science begins with the assumption of the Big Bang, but forgets to tell us about the origin of the matter from which it occurred, what caused the Big Bang, and why the Universe is so intelligently arranged. Science assumes the evolution of life, but does not explain it because the millions of species of the Animal Kingdom prevent a suitable explanation. The fundamental thing about the age of the Universe is that it was here before God created man. And can man limit God by mere thinking, when He has many options and infinite power.

God Brought Matter into Being by the Omnipotence of His Word

The spatial Universe was subject to a condition of heat – matter in form of gas. Job 38:37-38 seems to indicate that from the heat gas resulted and from the gas dust resulted. The dust then grew into hardness. This is indeed a description of the formative state of the Universe.

Figure 2.1

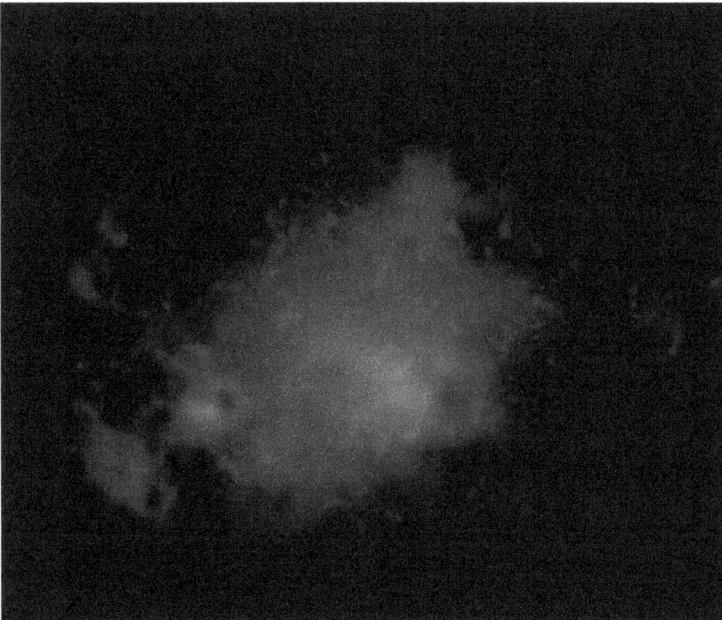

The Cooling and contracting of gaseous matter

Figure 2.2

Matter contracted into cold invisible matter beneath and clouds of dust and gas above. The cold invisible matter created the condition of weightlessness. The clouds of dust and gas, the Cosmic Cream, floated on. The formative stage of the Universe did not have to expand as scientists claim; it was absolutely miraculous – Omnipotence was at work..

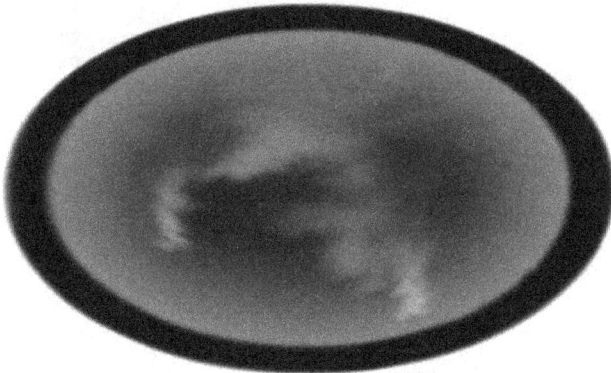

Footnote: The Theo-Cosmos Theory is founded on the interpretation of those Scripture references that address the nature of God and Cosmology. It becomes necessary to herein cite some of those references: the infinity of God, not limited to time and space. He is The Eternal God (Job 36:26; Isa.43:10,13;57:15; 57:15. He is omnipresent (Isa. 40:28; Jer. 23:24). He is clothed in majestic Light (Psalm 104:2; Job 37:18). He spoke matter into existence (Psalm 33:6-9;148:1-5)

-

<u>Galaxies of Stars Formed</u>

<u>Figure 2.3</u>
God spoke and separated large volume of matter into galaxies.

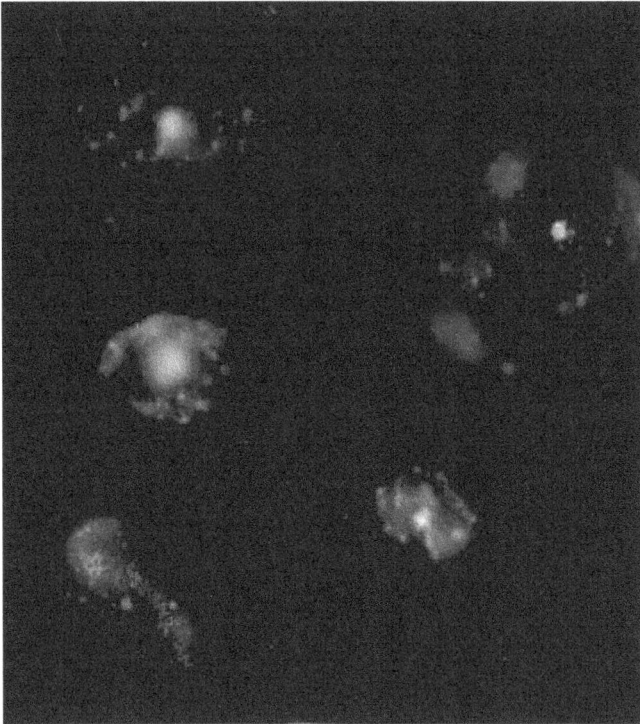

A Galaxy of stars

Figure 2.4

God put the laws of Gravity and Motion to work. Star formation began at the center at great velocity. These stars continue in their orbits and rotations with the velocities in which they were formed. Stars at the center of a Galaxy would be naturally older than other stars.

The Theory of the Evolution of the Solar System

Scientists believe that the Solar System was once a cloud of dust and gas. They believe that this cloud of dust and gas was the result of a star explosion 5 billion years ago. They claim the Sun is a third generation star. According to them, an earlier generation of stars was necessary to create the heavier element of the present generation of stars.

There are different theories of the origin of the Solar System but are closely similar in substance. The following is a paraphrase of a theory proposed 1949 by G.P. Kuiper: The Sun was formed by the contraction of a rotating cloud of gas. Some of the matter escaped the process of formation of the Sun. While the Sun was forming, the matter that escaped the formation process broke up into smaller volumes and formed into nine planets. Kuiper assumes that the Sun and the planets were formed simultaneously and that when the Sun started to rotate large portions of the outer masses of the planets were blown into space because of their high temperatures. Some of their masses were also lost to evaporation. <u>Atlas of the Universe by Thomas Nelson and Sons,1961 Ed, p102.</u>

Our response to this theory is similar to our response to the Big Bang Theory with regards to the duration of the process of formation as the basis for calculating the age of the Solar System. As we have stated in the THEO-COSMOS THEORY that the Universe was once in a state of dust and gas in its early formation, we accept that part of the theory to be authentic. While we accept that parts of the Universe are older than other parts, we refuse to accept that the age gap between any two parts could be 5 to 10 billion years, or that the Sun is a second or third generation star. We take the position that the Solar System was formed in the beginning with the rest of the Universe from gas of clouds and dust; but the conditions for life and life forms were not created until sometime after.

The Theory of the Development of the Earth

The development of the Earth is interestingly explained by the geologists. They explain that the present form and condition of the Earth took a period of approximately 5 billion years. They divide this time into six geological periods. We do not know if these 6 periods were intended to correspond with the 6 creative days of Genesis chapter 1.

The First Geological Era—Archeozoic. This era lasted 1, 500, 000, 000 years. During this period the Earth began its existence as a body of hot liquid and gas. As the hot liquid cooled, it formed into solid rocks. In the process a thin crust began forming into the Earth's surface. This crust often broke up as more liquid rock spilled on. The weight of the liquid rock caused the crust to sink in some areas and to rise in other areas. While the Earth continued cooling, condensation in the atmosphere began in the form of rain. It rained for millions of years until the Earth became cool enough, but the Earth's center remained a hot dense liquid.

After the Earth's crust was formed, parts of it were broken up by thousands of years of rains and winds. Rocks were broken into small particles and carried thousands of square miles on the Earth's surface. This era ended with the beginning of life.

The Second Geological Era—Proterozoic. This era lasted 1,500, 000,000 years. It saw the development of plant and animal life in the oceans. Later animals began to live on land. Further changes in the Earth's surface began to occur through volcanic activities, causing parts of the crust to merge while other parts plunged beneath the seas. Many mineral deposits were formed by the cooling process of the lava.

The Third Geological Era—Paleozoic. This era lasted 800 million years. A major feature of this era was that plant and animal life became more complicated and abundant. Sea animals began moving out of the waters to live on land. Many types of insects developed. Great coal fields formed from buried plants and trees. Further changes in the Earth's crust occurred. Great layers of ice formed at the North and South Poles. Winds and rains effected further changes in the Earth's surface.

The Forth Geological Era——Mesozoic. This era lasted about 800 million years. Remarkable biological changes took places within many animals. Animals began to develop hard bones. Many amphibious animals began to appear. The reptiles became the dominant animals of this period. Noted also was the appearance of fruit trees.

The Fifth Geological Era——Cenozoic. It lasted about 399 million years. Very early, warm blooded animals replaced the reptiles. Dominant among them were the mammals. Elephants and horses were among the first to appear. In the mean-time, changes continued to occur in the Earth's surface.

The Six Geological Era. This was the shortest era. It lasted about a million years. At the beginning man appeared on the scene but not in his present form. His present form took thousands of years of development. Noticeable in this period was the covering of the Earth's surface with ice. The Earth is said to have been covered with ice from the North to the South Pole four times.

The scientific theories of the origin of the Earth and its development are not theories we can completely reject. Without hesitation, we can emphatically reject the accidental aspects of the theories with respect to the origin of the Earth and the evolution of life. But we have no Biblical ground to reject the age of the Earth and its stages of development. We would question the durations in the stages of development in the plant and animal kingdoms. But we cannot condemn or reject the stages in the Earth's development as explained by the geologists since the Biblical Account begins with an Earth already formed except for the conditions of life and life forms. The Bible does not tell us how God formed the Earth; it only gives some details of the creation of life conditions and life forms.

From the Genesis Account of Creation, one could assume that God created the Universe progressively. Yet such assumption is neither evidence nor proof. Theology is not based on assumptions; it is based on faith, revelation, and on experience. It does not attempt to prove God's existence. It declares it and cites the Universe as the evidence. A finite being could not prove the existence of an Infinite Being, no matter how he tries. That was a job left up to God. Quite frankly, He has proved His existence countless times. He has proved His existence to Israel beyond a reasonable doubt and to Christians everywhere. This gives Theology the surest foundation it could have ever hoped to have. The Universe was created before there was any matter in existence; this is the meaning of Creation. Let this challenge our feeble minds and bring us face to face with the reality of God and the wonders of His works.

The Theory of the Evolution of Life

The Evolution of Life Theory is an under-developed theory. It is fair to say the Theory itself is not fully evolved and there is no likely time in sight. The question of life that it must answer remained unanswered. The origin of life remains unexplained. The chemical properties of life can be identified. Some of these chemical properties are carbon, oxygen, hydrogen, nitrogen, iron, and calcium. They are found in matter around us. To date, scientists have failed to put them together in a manner to produce life. Scientists say all forms of life evolved from a living cell. Yet they have tried and failed to explain how that living cell came about.

There are many versions of the Theory. The one that will be here considered is the version Hazen and Trefil present in their book, Science Matters. After stating the renowned premise of the Theory that all life forms evolved from earlier and simpler forms, they argue a two-step process: Chemical Evolution and the Biological Evolution.

The Chemical Evolution was critical to the Biological Evolution and so naturally preceded it. In the Chemical Evolution of life, the oceans became indispensable. They were the mixing bowl. The Earth's early Atmosphere played an equally important role in the Chemical Evolution. It provided such elements as carbon, hydrogen, oxygen, phosphorus and sulfur. These gases mixed with the surface layers of the oceans. This mixture formed the complex molecules necessary for life.

Hazen and Trefil argue that…1953 Stanley Miller and Harnold Urey at the University of Chicago conducted an experiment to determine what natural process could produce the complex molecules necessary for life. The results of the experiment revealed amino acid is the building block for proteins. The longer the experiment lasted, the more diverse and concentrated the matter in the experiment became. It was to the degree that some people had thought new and dangerous forms of life might emerge from the test tube.

Hazen and Trefil further argue the Chemical Evolution of life by referring to the "Primordial Soup". This is the Theory that the radiation from the Sun and lightning energized and combined the simple gases in the upper layers of the early oceans into complex carbon based molecules. The process involved millions of years. However Hazen and Trefil have no definite conclusions as to how and when the first living cell originated. "We do not know how life arose from the Primordial Soup. This remains the greatest gap in our knowledge." Science Matters op cit. p.247

Once the Chemical Evolution occurred, the Biological began. Unlike the Chemical Evolution, which had taken millions of years to produce the first living cell, the Biological Evolution took a brief time in comparison. In a short period of time that first living cell multiplied itself to the degree that it filled the oceans. The Theory concludes that the diversities of all life forms on Earth evolved from that first cell by Natural Selection.

RESPONSE: This part of the Theory is a gross misinterpretation of Darwin's Theory of Natural Selection, which focus is on the survival of a species. The Evolution of Life Theory is unable to answer at least four questions of life. First, it has failed to explain exactly when and how life began. Expressing an idea that the first living cell may have taken millions of years to form is not really telling us when life originated. And identifying the chemical elements of life is not really telling us how life was formed or evolved. The fact that evolutionists are unable to put the chemical elements together to create life is the evidence that the theory has not explained when and how life began.

Second, the theory cannot account for the millions of species in the animal kingdom. If life evolved from a single living cell, how came these species of such great diversities? Some evolutionists may have misused Charles Darwin's doctrine of Natural Selection to explain the species of the animal kingdom. But the true meaning of Natural Selection lies buried in certain differences within a species. Looking at the Doctrine of Natural Selection for a brief moment, it will be obvious that Natural Selection perpetuates a species and says nothing about how a species originated. Look at the premise in the words of Darwin: "The preservation of favorable individual differences and variations, and the destruction of those which are injurious, I have called Natural Selection, or the Survival of the Fittest". The Origin of the Species by Charles Darwin, University of Oxford Press, Sixth Ed.p181. It is clear that Natural Selection deals with the preservation of species and not the origin .

Third, the Theory is completely silent on the most vital question of life, the question of intelligence. In the plant kingdom the characteristic of intelligence is absent. This is what to be expected from accidental evolution. In the animal kingdom the situation is

Footnote: The Scientific theory of the Earth's development and the Theological response herein expressed are direct excerpts from my work, Understanding God's Sovereignty.

completely different. Why should there be the various levels of intelligence as they are? Can the Theory explain the staggering degree of human intelligence? Certainly not!!

Forth, the Theory does not answer the most fundamental question of life, the question of the moral and spiritual nature of man. The fact of man's moral and spiritual nature is a fact taken for granted and many times greatly understated. But the recognition of man's moral and spiritual nature is universal, and has impacted the governments of the World. All governments may not be the Western Style Democracy we know. Nevertheless, they are all based on the moral code of man's nature. In their own way, they all seek the good of their respective citizens. The spiritual nature of man is from the fact that men throughout the World have demonstrated an inherent desire for worship. Men everywhere have an intuitive knowledge of a superior Being or beings to whom they owe their existence and to whom they feel a kinship, a mutual sense of love, a confidence, and hope of one day comprehending more fully the knowledge of God. Strange enough, religion as we know it, predated human governments: so fundamental is the spiritual nature of man. The Theory has failed to answer these four fundamental questions of life. We are left no choice except to reject it in its entirety, unequivocally.

Science has tenaciously refused to take its rightful place in the intelligent Universe as a mere complementary role to Theology in explaining nature. Science cites its observations in support of its theories and deductions. In like manner religion can cite the many miracles of the Bible in support of its claims of the existence of God. The Jews of today are living evidence to those claims. Millions of people today have had personal experience with God, including this Author. Science can always ignore or deny these realities but they are no less real. The denying or affirming of a fact does not change it. The denying or affirming of the moral and spiritual code which underlies man's inner nature will not change the reality of it. Those scientists who disregard the Bible and the God of the Bible still do have a moral and spiritual nature. This they cannot change, but they can change their choice of not living out the potentials of their moral and spiritual nature.

CHAPTER 3

Scientific Theories: Review and Response Continued

Some scientific theories can be most valuable. On the other hand, some can be most uninspiring. What can be uninspiring about a scientific theory is not its face value, but its implications. One can read into a theory what was never intended, and also the unwillingness of scientists to limit the implications of a theory. Einstein did this when he had introduced the Cosmological Constant to balance the attraction of gravity. The same thing can be said of Quantum Mechanics. To apply it to the Universe to say that it is governed by chance is, indeed, the most dangerous implication of Quantum Mechanics. In this chapter we want to review some of the better known theories. They are the theory of the Expanding Universe, Newton Mechanics, Relativity, and Quantum Mechanics. These theories suggest the curiosity of the human mind and its infinite capacity for knowledge.

The Theory of the Expanding Universe
Some scientists believe the Universe is expanding. The Hot Big

Bang Theory is essentially about the Expanding Universe. The rate of expansion immediately after the explosion was incredibly great. With the passing of time billions of years after, the rate of expansion greatly reduced. Today the Universe continues to expand but just at the critical rate to balance the attraction of Gravity. A static Universe would have caused it to collapse or disintegrate. The Expanding Universe is the subject of the third chapter in the book, A Brief History of Time written by Stephen Hawking a theoretical physicist. In the third chapter he emphasizes the spatial expansion of the Universe. The distances between galaxies are increasing constantly. The Hubble Discovery in the 1920's showed that the galaxies at the outer reaches of the Universe were moving away and the farther away, the faster they were moving. This discovery forms the foundation of Stephen Hawking's argument for the Expanding Universe. He argues three points to support the theory of the Expanding Universe. They are the Doppler Effect, the Common Similarity with parts of the Universe, and the Microwaves Radiation.

The Doppler Effect is applicable to sound as well as radiation. In the case of sound, the frequency is higher moving towards its source, depending on the speed at which one moves. The frequency or sound is lower, moving away from the source, depending also on the speed at which one moves. In the case of radiation, measuring the distance of a star, a shift to the blue end of the spectrum means that the star is moving towards the measuring instrument. A shift to the red end means that the star is moving away. Accordingly, all the galaxies in Hubble's discovery showed a red shift: the galaxies were moving away, meaning the Universe is expanding.

Response to the Doppler Effect. We are not questioning the principle of the Doppler Effect. We will offer three considerations which will articulate it but will contradict the theory of the Expanding Universe.

First, is the uniformed directional motions of bodies in space. We know that the Earth and the other planets of the Solar System move west to east clockwise, following the motion of the Sun. One could assume this directional motion is true of other motions in the Universe. Hubble's discovery that the galaxies at the outer reaches of the Universe were moving away from each other does not contradict the Hypothesis of the Rotation of the Galaxies: the stars are orbiting within the galaxies. A great many stars are seen orbiting one another. These stars could not simultaneously orbit the core of the Galaxy. Notwithstanding, its Gravity holds all its stars together.

The orbits of stars are greater in distance than the orbits of the planets of the Sun. With this consideration, the angles of these stars cannot be ignored. These orbits will take many years to complete. And if seen at different angles moving to and from each other may show a Red or Blue Shift. An orbiting body moving at an angle of 180 degrees towards the measuring instrument will show a Blue Shift, quite possible for many years, depending on the distance of the orbit. But first, the astronomer must know the distance of the orbit

The infinity of the Universe will always present problems for us in our understanding of its behavior. But we must try to balance interpretation with reality. If the Milky Way should be viewed from one of those receding galaxies, there would certainly be a red shift, even though scientists inform us the Milky Way is rotating on its axis. Looking from one of these receding galaxies, astronomers would see the galaxies at the center of the Universe spreading apart, though they are not now thus seen. The reason is that the receding galaxies would widen the space between themselves and the galaxies at the center of the universe and thus show a red shift.

Perhaps the interplay between distance and motions in the infinity of the Universe has played a trick on our eyes. Perhaps the red shift of the outer galaxies merely reflects the total motions in the Universe, since all the galaxies are in perpetual motions. Why is it that when we look up we see a seemingly impenetrable mass – the blue sky? This view may have been caused by the Earth's atmosphere and the reflection of the Sun's radiation on the oceans.

The masses of the Solar System have warped space as well as all other masses in the Universe. Looking from any of the planets of the Solar System, one will see a sky above, the Cosmic Umbrella. This is due largely to the round shape of the planets. Should one look at the sky from deep space, the shape of the sky as seen from the Earth, would look completely different.

The authors of Science Matter, Hazen and Trefil express great confidence in the Theory of the Expanding Universe. Also, like Stephen Hawking, their argument supports a spatial expansion, the mere increase in distances between the galaxies. They use the analogy of the Raisins and the Bread Dough. The dough represents space; the raisins represent the galaxies. The raisins at the edge of the dough are receding faster than the other raisins because the dough (space) is expanding. Stephen Hawking, Hazen, and Trefil have not been singled out for a "preemptive strike" against their support of the Theory, but because they might not admit that kind of expansion implies an intelligent cause. If I be generous to say there is a spatial expansion of the Universe, they must be generous to say such an expansion implies an Infinite and Intelligent Cause.

The Common Similarity with Different Parts of the Universe
This is the second point Hawking uses in support of the Theory of the Expanding Universe. This point has been based on the assumption of Russian physicist, Alexander Friedman. "Friedman made two very simple assumptions about the Universe: that it looks identical in whichever direction we look, and this would also be true if we were observing the Universe from anywhere else. From these two ideas alone, Friedman showed that we should not expect the Universe to be static". A Brief History of Time, Bantam Books, 1990 Ed., p. 40

We are partially questioning the validity of Hawking's interpretation of Friedman's assumption. While we agree that in a large scale, the Universe would look well similar; we completely disagree on the cause – the expansion of the Universe. As previously stated, we believe that the Universe is filled with invisible matter and the solid masses have warped space. This would cause the Universe to look similar. This warping of space time has created the Cosmic Umbrella. However, varying locations from which the Universe is viewed will show variations in how the Universe looks. We see a blue sky above us, but it is due to the location of the Earth from which we view it.

The Microwaves Radiation. This is the third point Hawking argues in support of the expanding Universe. During 1964, American scientists at the Bell Telephone laboratories, Arno Penzias and Robert Wilson were testing their microwave detector. They detected unusual noise in their detector. They found that the noise was coming from all directions the detector was pointed. They concluded the unusual noise was coming from outside the Solar System and from beyond the galaxy of the Milky Way. They thought the radiation from the hot early Universe was now reaching us but the expansion of the Universe caused it to have appeared as microwave radiation.

He further explains that scientists Bob Dike and Jim Peeble of Princeton University were preparing to carry out experiment on Microwaves Radiation when they heard of the discovery of Penzias and Wilson. Dike and Peeble were studying George Gamow's idea that the early Universe was very hot and glowing. Dike and Peeble reasoned that we should still be able to see the glow of the early Universe, but only as Microwave Radiation (electromagnetic waves of high frequency) due to time and distance.

Hawking concludes that the expansion of the Universe would cause the glow of the early Universe to now be seen as Microwave Radiation. He estimates the rate of expansion to be between 5 to 10 percent every billion years

Response: Microwave Radiation is a reality of nature; the problem is with its cause. I am wondering why scientists have never thought that the Microwaves Radiation could be the result of the radiation of all the stars which due to time and distance has settled down to the form of Microwaves Radiation. The Universal Law of the Conservation of Energy that energy can never be created nor destroyed implies that the radiation of the stars must remain in the Universe in some form. Microwave Radiation could only be a more stable form of the radiation of the stars.

The Premise of Theory: If the Universe were not expanding it would have already begun to collapse. The expansion is necessary to balance the attraction of Gravity. The Universe is not merely expanding but must expand at certain rate. Again I quote Stephen Hawking: "The discovery that the Universe is expanding was one of the great intellectual revolutions of the Twentieth Century … Newton and others should have realized that a static Universe would soon start to contract under the influence of gravity." A Brief History of Time I bid..p39

Response: What this premise is in essence is a mere scientific idea which is very much remote from reality. In our Solar System the nine planets orbit the Sun at comparatively close distance when compared with other distances of outer space. The principle in the Solar System is that the velocity of the motions of the planets balances the attraction of the Sun's gravity. Because of the motions of the planets they do not crash into the Sun. This principle is precise. The rotations and orbits of the planets remain the same year after year, decade after decade, Millennium after Millennium. By this principle the positions of the planets can be always predicted with perfect accuracy. The Earth has a velocity of rotation of 1000 miles per hour and orbital velocity of 66,000 mph. This allows for an accurate prediction of the Earth's position at any time.

Stephen Hawking is well aware of the principle of velocity of motion balancing the attraction of Gravity. He explains that the orbit of the Earth around the Sun produces gravitational waves (loss of energy in motion) and that the end result of the Earth's gravitational waves will be that it will collide with the Sun. But he points out that this would not happen before a thousand, million, million, million-million, million years. "The rate of energy loss in the case of the earth and the Sun is very low, about enough to run a small electric heater. This means it will take about a thousand million, million, million-million, million years for the earth to run into the Sun, so there is no immediate cause for worry." A <u>Brief History of Time,</u> <u>Ibid p.90</u>

A principle is only a principle if it can be applied without restrictions. The principle of Velocity of Motion Balancing the Attraction of Gravity, with the modification of gravitational waves, (loss of energy due to motion) must also be applied to all the stars in the Universe. When thus applied, many scientists will be more than happy to abandon the premise of the Theory of the Expanding Universe. WHY? Because when applied to distances (unlike the distances of the Solar System) of many light years between stars, it would take, not years, many eternities before one star would collide with another star.

We must on the ground of the principle of Velocity of Motion Balancing the Attraction of Gravity and on the ground of Hawking's own argument on Gravitational Waves, reject his claim that the Universe is expanding to balance the attraction of gravity. Since his argument shows it would take countless years before the Universe collapses, we can be confident in emphasizing that if the Creator so desires, the Universe can exist forever in its present form.

General Response to the Theory of the Expanding Universe

There are many things we do not know about the Universe. And that which we know is characterized by much uncertainty. But there is no imperfection, nor uncertainty with the laws of nature. They are constant, not changed by time and conditions. Rather, time and conditions are determined by the laws of nature. The discovery of a new phenomenon in nature does not necessarily mean the discovery of a new law of nature. But it does call for a closer examination of the known laws. Indeed, it is this closer examination that will determine whether a new law has been discovered. Until that is done, we should expect that the already known laws do apply. Thus we expect that the laws that govern the galaxy of the Milky Way also govern the galaxies of the Hubble's discovery. Our examination of these known laws and their application must be thorough. Is there something we have omitted? We must search diligently.

Many years ago, when I bought my first car, I took it to a mechanic for engine repairs. I wanted to learn something about the car, so I started a conversation with the mechanic. I do not remember the details of that conversation. But now I clearly remember four words the mechanic used——the "Centripetal Force" and the "Centrifugal Force". Although these words were not new to me, I was, however, surprised hearing them applied to the motions of the engine. The intake valves represent the Centripetal Force; the exhaust valves represent the Centrifugal Force. The two forces counterbalance each other and keep the engine in motion.

We do not know our position in the Universe, whether we are at its center or otherwise located. We do not even know our position in our galaxy of the Milky Way, even though it is believed our Solar System is located fairly close its center. However, our lack of knowledge places no limitation on the application of the principles of the Centripetal and the Centrifugal Force to the motions in the Universe. The laws of nature were established during its formative process.

These two great forces sustain the motions in the Universe. The Centripetal Force makes motions inclined in ward, moving towards. The Centrifugal Force makes motions move outward, away from. These two great forces are inwrought in the very fabric of the Universe and in all of nature. In any mass of matter there is the tendency to pull towards its center, and there is also the tendency to pull from its center. At the center of the Universe the Centripetal Force is strongest; the Centrifugal Force is weakest. This slows down the motions of the galaxies at the center of the Universe. At the outer reaches of the Universe the Centrifugal Force is strongest; motions are fastest.

Our concept of the center of the Universe is that region that has the largest number of galaxies when compared with any other region, or combinations of regions. The greater configuration of mass, the greater the slowing down effect of the Centripetal Force: galaxies at the center move slower. The farther away a galaxy is from the center, the faster is its speed. Velocities of motions do not contradict the known laws of science. Thus the faster motions of galaxies at the outer reaches of the Universe are not evidences of an expanding Universe. An evidence of an expanding Universe would be an increase in distances among the stars in the Milky Way and noticeable formation of new stars between the expanded spaces. Only a substantial expansion, one in which both volume of matter and distances increase could be considered an expansion. The scientists have not informed us of such an expansion.

God may have created an infinite Universe in the initial act of Creation in the Dateless Past. Though it would have involved an indefinite period of time, it would not necessarily have an expansion aspect. This point can be illustrated by the analogy of a building and its builder. A very large building may have been completed in a non-stopped construction. On the other hand, it may have been completed by additional constructions at different times, like a poor man who first built a two bedroom house and later built additional rooms. In this case we have favored the former as the way God constructed the Universe. If there were any additional expansions as the latter, that expansion ended the day God created the life conditions and life forms of the Earth. What we know about Creation is only what has been revealed

I find great delight both in science and in the Bible that the

Universe is measureless. I find great delight in this fact because it is the strongest proof of the infinity and the omnipresent nature of God. Thousands of years ago, before the dawn of modern science, the Creator declared through the prophet Isaiah: "Ye shall not be ashamed nor confounded world without end" (Isaiah 45:17). Again He declared through the Prophet Jeremiah: "The host of heaven (stars) cannot be numbered." (Jeremiah 33:22). It is obvious that mankind is only now discovering the infinity of the Universe. We must understand that the essential nature of a discovery is to reveal the character of a thing, not to change it. Discovering the **infinity** of the Universe is quite a different thing than the assumption that it is expanding.

Theories of Gravity and Motion (Newton Mechanics)

The Theories of Gravity and Motion otherwise called Classic Mechanics were formulated by the brilliant English scientist, mathematician, physicist, and astronomer, Sir Isaac Newton (1642–1727). Newton's theories were first published in 1687. Newton was preoccupied with the motions in the Universe, particularly those of celestial bodies. He saw that the motions of these bodies were regular and systematic. He had been considering other motions too. He knew that motions were not spontaneous. To get a stone moving would require an external force. The larger the stone, the greater would be the force required to get it moving. There was always a tendency of resistance to a change of position. A stone rolling down a hill will continue rolling unless it is stopped by an external force. This tendency to resist a change in position, Newton calls "inertia" Newton found that all masses have a degree of inertia, which can be overcome. To overcome the inertia of a mass means a change in position of that mass. To Newton there were only two basic kinds of motions: uniform motion and accelerated motions. Uniform motion is continuous motion without any change in speed. Accelerated motion is a motion in which there is a change of speed.

Newton sums up the motions of the Universe in the following three laws:

1.A body at rest will remain at rest, and a body in motion will continue to move in a straight line at the same speed, unless the body is acted upon by an outside force.

2.Acceleration, or change in velocity is directly proportional to the force and inversely proportional to the mass. It is evident that this law deals primarily with motions on Earth since celestial bodies are in a uniform state of motion. The motions of motor vehicles best illustrate this law. The greater the force is, the greater the acceleration, direct relation. The greater the mass is the slower the acceleration. A heavy motor vehicle will move off more slowly than a light one with equal force, inverse relation.

3.Every action causes an equal and opposite reaction. When you push against an object that object pushes against you with equal force. The force of the push is always balanced. This principle applies equally to motions on Earth as well as celestial motions. The gravitation force exerted by the Sun on the planets is in turn exerted by the planets on the Sun. Because of the greater mass of the Sun (98 percent of the matter in the Solar System) the planets orbit the Sun. In mechanical engineering, the centripetal force of the motor engine is balanced by the centrifugal force.

GRAVITY. As an astronomer Newton knew that the celestial bodies move in circular line. He reasoned that there was some force that causes this to happen. He called this force Gravity. Newton's concept of Gravity was conceived when he saw an apple fall from a tree while he was able to see the Moon. He reasoned that the Earth's Gravity which forced the apple to the ground could have extended to the Moon and caused it to circle the Earth instead of falling to the Earth, due to the distance between the two. After much deliberation, he articulated three laws of Gravity known as The Universal Law of Gravitation. They are as follows:

1 Everybody attracts every other body in the Universe. This law implies that a body will orbit the closest center of gravity.

2. The amount of attraction is directly proportional to mass. The size of the mass determines its gravitational force, or which of the masses in a group is the center of gravity. In the case of the Solar System, the Sun is the center of gravity.

3 The amount of attraction is inversely proportional to distance. The farther the distance between two bodies, the less the force of gravity is felt, the one body on the other. Because of the relatively closer distance of the Moon (240,000 miles) than the Sun (93,000,000 miles), the Moon's Gravity is stronger on the ocean's tides than the Sun's. There is no limit to the application of Gravity; it can be applied at the level of the atom as well as to celestial masses.

RESPONSE: Newton Mechanics is not a perfect theory: it is incomplete. It had three major deficiencies. It did not describe the nature force of Gravity. It did not suggest a medium through which Gravity works. Neither did he suggest how rotations and orbits were established. He did not explain that the force of Gravity was magnetic attracting force and how exactly Gravity works over long distances. Newton Mechanics only explains what happens, once the mechanism of the Universe is switched on. He himself recognized this deficiency and was often questioned about it. He never seemed to have had the explanation. "To us it is enough that Gravity does really exist, and act according to the laws which have been explained, and abundantly serves to account for all motions of celestial bodies and our seas." Encyclopedia of Philosophy by Cromwell and Mac Milan Inc., 1967 Ed. V1. 5. P 490.

Many other scientists were sympathetic towards Newton's apparent failure to explain the nature of Gravity and how it works over great distances and suggested a contact medium of micro-physical entities and processes. Newton did not object to this idea. In-spite of this apparent weakness in the Theory, Gravity is an in-escapable reality of nature, permeating all physical phenomena as well as the simple everyday things of Earth. Without the fact of Gravity we would not have had an adequate definition of the measure of weight. Weight is the pull of Gravity on a standard object with the pull of Gravity on the object being weighed.

During the course of writing Understanding God's Sovereignty, I had sensed that I have had a special intuition for science. I then began to think about some unanswered questions in science. After writing that book, I felt my job was not yet completed as I was not able to deal with some of the questions that were going through my mind. One of these questions had to do with Gravity – how Gravity works. Before I had begun writing THEO-COSMOS and before I had done any research, the answers to these scientific questions were already clear in my mind. I now remember saying in my book, Understanding God's Sovereignty that the electromagnetic properties of matter are a factor in the attraction of Gravity. I knew then that this was not the full answer to the question: it was only part of the answer. After thinking of the question, I later saw a similarity with the attraction of a piece of natural magnet and the attraction of Gravity.

If a piece of wire or metallic object is brought within a certain proximity to a piece of magnet, it will attract the wire unto itself. A magnetic-field is created around the magnet and within its immediate area. A magnetic field is also created around the piece of wire. These two magnetic-fields interact. The result is that the wire is attracted unto the magnet. We can deduce that the wire and the magnet have much in common and that the basic difference between the two is that the magnet has more magnetism than the wire. The two principles in this mysterious happening are the similarity in matter and the magnetic fields. The action of Gravity must work by these two principles. Matter is basically composed of electromagnetic properties. This is particularly obvious of celestial bodies (stars).

Radiation of stars is electromagnetic in nature. This radiation creates and sustains the Magnetic-field of the Universe. The Magnetic-field of the Universe works in conjunction with the magnetic-fields of the attracting mass and the attracted mass. The universal law of the Conservation of Energy (energy cannot be created or destroyed) dictates that the radiation of stars must remain in the Universe in some form. Have we ever asked ourselves, "What has happened to all that energy generated by the stars?" Well most of that energy sustains the Magnetic-field of the Universe through which all gravitational activities occur. In a subsequent chapter more will be said about cosmic radiation (radiation of stars).

One cannot separate the Magnetic-field of the Universe from the attraction of Gravity. They are critical to the motions in the Universe. The evolutionists will say if that be the case, the Universe would have collapsed many times over. This would be true if the Universe had accidentally evolved. And still this would not be an over-night event because the attraction of Gravity is balanced by the velocity of motions. Newton Mechanics did not attempt to explain how rotations and orbits were established. It explains and predicts what happens and will happen after the mechanism of the Universe is switched on. In the Theo-Cosmos Theory of the formation of the Universe, it was pointed out that velocities of motions were preexistent in the formation process of bodies in outer space. We can take this aspect of the Theo-Cosmos theory to complete the Newton Mechanics. Our experience with the Solar System has clearly established the fact that velocities of motions were predetermined by the process of formation of the Universe. If we put a satellite into orbit at a certain distance from the Earth, it becomes an independent satellite of the Earth. And it will continue to orbit the Earth at the same distance and speed at which it was put into orbit. If it was put into orbit at 3,000 mph. speed of rotation, 8,000 mph. speed of orbit, and at a distance of 5,000, miles above the Earth, it would thus continue to orbit the Earth..

While Newton Mechanics did not show how velocities of motions and orbits, in the case of the Solar System, were established, it graphically shows that velocity of motion balances the attraction of Gravity. If the planets were not moving as they do, they would have eventually crashed into the mass of the Sun. With these constructive suggestions, Newton Mechanics will not be superseded.

In concluding a word must be said with regards to Newton's interpretation of time and space. Newton believed in the absolute nature of time and space: time and space are not affected by events in the Universe. Time and space effect changes but they themselves are not affected by those changes. In my opinion this belief is justified when considered that the Universe was created by God. The Universe had a definite time of beginning. All events are therefore directly and indirectly associated with that point of beginning (God's time). On the other hand, Earth has a standard time. All times can be measured by a particular standard of time in which case the absolute nature of time is preserved. As will be pointed out when we come to the theory of Relativity, space time is only a new awareness of time.

The Theory of Relativity

The Theory of Relativity is synonymous with the name Albert Einstein (1879-1955). Einstein gained world fame for the theory due in part to his scientific brilliance expressed in one aspect of the Theory and in part to the controversial nature of another aspect. His scientific brilliance was obvious in Special Relativity published 1905. The controversial aspect sprang from General Relativity published 1915.

The Theory of Relativity in a way is most deceptive as its meaning is not clearly expressed of the obvious relativity between one thing and another in the World of nature. From a Theological point of view, one would think that the Universe is the work of one Omniscient Mind and the relativity in nature is confirmation of this truth. This Theological expectation of the Theory has been disappointing. The Theory in its essential nature probes deeply into the Cosmos, makes observations, and utters predictions of the nature and behavior of matter. This is particularly true of Celestial Bodies and their motions.

The complex mathematical equations presented in the argument of the Theory make it difficult to be understood. Surprisingly, other scientists who have understood the Theory have explained it with absolute clarity, yet without the mathematical equations of its original presentation. We can sum up the Theory thus: Pre-Relativity Physics. Pre-Relativity physics explains the nature of time. It also explains the principle of Relativity with respect to direction. In other words Einstein began his Theory of Relativity by the law of Apperception.

This law of Apperception is the method by which one teaches something new by starting from that which is known. On the other hand, Pre-Relativity physics could be seen as a subtle attempt aimed at the very foundation of Newton Mechanics (theories of Gravity and Motion). "In the first place it is assumed that one can move an ideal rigid body in an arbitrary manner. In the second place, it is assumed that behavior of ideal rigid bodies towards orientation is independent of the material of the bodies and their changes of position, in the sense that if two intervals can once be brought into coincidence, they can always and everywhere be brought into coincidence." The Meaning of Relativity published by Princeton University Press, N.J., Fifth Ed.

Special Relativity. The Michelson-Morley Experiment of 1887 was the foundation of Special Relativity. This experiment abolished the Theory of the Ether. The Theory of the Ether, in its essential nature, was that an undetermined substance permeated the Universe solely for the purpose of propagating light. Light needed a medium through which to manifest. The Ether was considered to be of absolute position, motionless. It should determine the motions of all bodies in the Universe. The experiment showed there was no Ether. It also showed the speed of light (186,000 miles per second) was constant irrespective of the speed of motion of an observer.

Michelson and Morley in their experiment used the Earth as an example. If the Earth is moving through the Ether, its motion should cause an Ether breeze and a beam of light traveling against the Ether breeze should have a slower velocity than a beam of light traveling across it, they reasoned. They invented a device called the Interferometer to conduct their experiment. It showed that there was no difference in the speeds of the two beams of light. However, these results were inconclusive because they were challenged by Irish scientist, George Francis Fitz-Gerald in 1892. A year later Fitz-Gerald gained the support of Dutch scientist Hendricks Anton Lorentz. This joint response to the results of the Michelson-Morley Experiment came to be known as the Fitz-Gerald Lorentz Contraction. The Fitz-Gerald Lorentz Contraction proposed the possibility of the arm of the Interferometer pointing into the Ether breeze being shortened, reducing the distance the beam of light would travel. This decrease in distance would be responsible for its faster speed and would compensate for the slower speed of light of the longer arm traveling against the Ether breeze. Both beams of light would, therefore, be recorded at the same speed. The important thing was that there was no Ether Breeze.

With the appearance of Special Relativity 1905, all inconclusiveness was removed from the results of the Michelson-Morley Experiment. So the results of the experiment became the premise of Special Relativity. The speed of light is constant and should be the same to all moving observers irrespective of their speeds of motions. Special Relativity introduces a new interpretation of time, Space time. Special Relativity shows how time is related to space. This relativity is obvious from the fact of an event that occurred at some point in space. For this particular event there are three observers at three different locations in space. Each observer has a perfectly accurate clock, but owing to each one's distance in space from the point of this event, each one's clock will record a different time and yet be correct.

The determining factors in the relativity of time and space are the speed of light and distance. Light travels at a speed of 186,000 miles per second. An event occurred in space. A ray of light is beamed from the event to three moving observers at three different locations from the event; 93,000,000 miles, 46,500,000 and 31,000,000 miles. The clock of one observer will show a time of 8 minutes; the clock of another will show a time of 4 minutes and the clock of the other will show a time of 2 and two thirds minutes. These three recorded times are correct but they have not changed the absolute nature of time because the determining factor of the speed of light was predetermined by some standard of time. We cannot think of weight nor length unless we have some standard of measurement. This equally applies to time. Time must have a standard implied or expressed. This makes time absolute. Space time is merely a new awareness of time.

Critical to Special Relativity is its equation of energy and mass. The Equation is spelled out as E=MC2, wherein E is energy, M is mass, and C is the speed of light. In practical terms, energy can be converted into mass and mass can be converted into energy. Energy equals mass multiplied by the speed of light squared. This equation tells you what potential energy is in a particular form of mass. Therefore, the energy a mass acquires through its motion increases its size. In the final analysis, when a mass is travelling close to the speed of light, its size doubles. However, it will never acquire the speed of light because as it gets closer and closer to the speed of light, its sizes continues to increase. This requires more and more energy, so it can never acquirer the speed of light. Therefore, the Equation limits all material objects to the speed of light: nothing can travel at the speed of light, except light itself.

General Relativity. It was obvious that Einstein had a thorough understanding of Newton Mechanics (theories of Gravity and Motion) and that in his mind his theory of Relativity had a clear and precise meaning. That this is so is evident from the fact that he proposed his famous Cosmological Constant, a relativity to balance the attraction of Gravity. The idea of an expanding Universe was not yet conceived. It was the general understanding that the Universe was static (not expanding). A non-expanding Universe would collapse under the attraction of Gravity if there was not a repelling force to balance the attraction, Einstein reasoned.

Einstein had deduced the concept of the Cosmological Constant from scientific ideas of fellow scientists, Maxwell and Poincare. "Matter consists of electrically charged particles. On the basis of Maxwell's theory, these cannot be conceived of as electromagnetic fields free from singularities. In order to be consistent with the facts, it is necessary to introduce energy terms, not contained in Maxwell's theory, so that the single electric particles may hold together in-spite of the mutual repulsions between their elements charged with electricity of one sign. For the sake of consistency with this fact, Poincaré has assumed a pressure to exist inside these particles which balances the electrostatic repulsion. It cannot, however, be asserted that this pressure vanishes outside the particles." <u>The meaning of Relativity, Ibid. p 106.</u> Herein was a picture of the eventual collapse of matter at the infinitesimal level of infinite density had it not been for a built in pressure or repelling force. This repelling force was inescapable. It was obvious in all of nature and in particular among the physical phenomena, preventing the Universe from collapsing under the attraction of Gravity, Einstein reasoned. With Edwin Hubble's discovery that galaxies at the outer reaches of the Universe were moving away at incredible speeds, the concept of an expanding Universe was conceived. It was then not expedient nor logical for Einstein to hold on to the relativity of his Cosmological Constant. He abandoned it, declaring it was the greatest mistake of his life.

We have earlier said that the Theory of Relativity probes deep into the Cosmos and makes predictions of the nature and behavior of matter. This is particularly evident in General Relativity. Einstein suggests that space and time affect changes in the Universe and such changes in turn affect space and time. "In the first place it is contrary to the mode of thinking in science to conceive of a thing (the space-time continuum) which acts itself but which cannot be acted upon." Ibid. p 56.

Now it is apparent from this statement that General Relativity predicted the expansion of the Universe. And other scientists who claimed that it did were dead right. It is clear that Einstein did not understand the full implications of his own statement that time and space are affected by events within the Universe. But the most controversial aspect of General Relativity is its interpretation of the motions of the Solar System and the velocity of Gravitational effect.

Stephen Hawking explains that the instant removal of an object that exerted Gravitational Attraction on another body would have infinite velocity because that object would be instantly affected. In such case, the velocity is greater than the speed of light. Einstein, therefore, suggested Gravity is not like other forces. That the motions of the planets are due to the fact that space is warped or carved by the mass of the Sun. The planets follow a straight path in carved space. Hazen and Trefil aptly present this aspect of the Theory thus: "For Einstein there were no forces in the Newtonian sense, only changes in the geometry of space. The relativistic interpretation of the Solar System, then, is that the Sun warps the space around it and that the planets move around in this space like marbles rolling around the inside of a bowl." Science Matters op. cit..p 168. From day one of the concept of Relativity, Einstein was opposed to Newton Mechanics and thus expresses his opposition. But he ingeniously disguised Newton Mechanics under the garment of Relativity and was thus able to make predictions of the same accuracy as predictions of Newton Mechanics. Einstein did not believe in the attraction of Gravity in the Newtonian sense, yet he explains that the mass of the Sun has so warped space that the planets move around the Sun.

The warping of the space by the mass of the Sun implies a tremendous gravitational attraction of the Sun in the Newtonian sense, which Einstein opposes but here unconsciously concedes. To Einstein the gravity of the Sun was potent enough to have made roads in space for the planets while it was not strong enough to attract the planets under conditions of the weightlessness of outer space. The point is that Einstein uses the attraction of Gravity, as we know it in the Newtonian sense for the warping of space and then makes the warping of space equivalent to the gravitational attraction of the Sun upon the planets. In this way and at all times the predictions of Relativity will be identical to those of Newton Mechanics. The warping or curvature of space has the built-in principle of Newton Mechanics of the attraction of the Sun's gravity upon the planets. Notice that the three major predictions of Relativity are the same and of the same accuracy as those of Newton Mechanics.

1. "The rate of a clock is accordingly slower the greater is the mass of the ponderable matter in its neighborhood." This prediction was confirmed by observation in 1962 with a pair of atomic clocks, one placed at the bottom of water tower, the other placed at the top.

2 "A ray of light passing near a large mass is deflected." This prediction was confirmed by observation in 1919. It was also confirmed by the "red shift" of the dwarf star which is physically associated with the large star, Sirius.

3 The Perihelion (nearest point when the planets come to the Sun) of the planet Mercury. Relativity attributes the changes in the perihelion of Mercury to the warping of space. All three predictions are also true of Newton's Mechanics.

Observations. What we know about matter is that it is made of neutrons, protons, and electrons. This means matter is electromagnetic in nature. This provides for gravitational attraction between two bodies, particular between radiant bodies like stars. The Magnetic-field of the Universe is filled with electromagnetic waves. This provides for the micro-physical entities and processes Newton Mechanics was once lacking. With this in mind, we understand that Gravity works over astronomical distances. Once an orbit is established, that orbiting body will continue its orbit forever.. Can Relativity and Newton Mechanics be both correct in explaining the motions of the celestial bodies? It may well be that none of these theories has explained the motions of the celestial bodies.

1 Every one knows that an object that is closer to the center of Gravity will experience a greater gravitational effect.

2. Many people who are not scientists know the five rules of light, one of which is that light can be deflected or bent.

3. Because the Sun is not a perfectly round body, at certain points its gravitational attraction will be greater than at other points. This will cause the planets to orbit closer at times. It is only common sense that Relativity and Newton Mechanics cannot both be right in explaining the motions of the Solar System, in particular, and all other motions in outer space in general.

The laws of Gravity must apply to all the phenomena in the Universe, or to none. No law of nature can be selective in its application.

The Theory of Quantum Mechanics

The theory of Quantum Mechanics is one of the many theories developed as a result of the study of radiation of light. It was proposed in the 1920's by German Scientist Werner Heisenberg (1901-1976). Like most theories, it has practical importance in understanding the nature of matter and in interpreting and understanding some laws of nature. Theories of the radiation of light led to the discovery of the speed of light (186,000 miles per second) between 1676 and 1865, and the concept of the Black Hole 1783.

Astronomers have made extensive studies of the radiation of light and thereby able to determine the chemical elements of the stars, their luminosities, temperatures, and distances. In truth and in fact the study of the radiation of light is the foundation of Astronomy. The microcosm of Quantum Mechanics was first seen in the Theory of Determinism proposed by French scientist Laplace at the beginning of the Nineteenth Century. Laplace is reported to have said that he swept the heavens with his telescope but could not find God. It was therefore not unnatural that he proposed the Theory of Determinism. The substance of Determinism is that a Supreme Intelligence possessing a knowledge of the Newtonian laws of nature and a knowledge of the positions and velocities of all particles in the Universe, at any moment could determine the state of the Universe at any other time. Without hesitation, many other scientists attacked Determinism. In their endeavor, a number of theories were developed. Quantum Mechanics was one of those theories blossomed and matured into fruits; a Theory recognized by many scientists as one the great theories of the Twentieth Century. While the microcosm of Quantum Mechanics could be clearly seen in the Theory of Determinism in that it was one of those theories developed as a result, Quantum Mechanics was a direct offspring of the Quantum Theory proposed by the German scientist Max Plank in the early Nineteenth Century.

Plank's Quantum Theory is significant because it has defined for us the Quantum World to the degree that it lets us know that radiation comes in discrete amounts called quanta. This gives us a panoramic view of the whole Quantum World— behavior of atoms and their particles. From the Quantum Theory, Werner Heisenberg developed the theory of Quantum Mechanics based on the Uncertainty Principle. Mechanics is the term used for the study of the motions in the Quantum World. But Heisenberg's Quantum Mechanics is equally significant because it has shown us the fundamental difference between our World and the Quantum World. This fundamental difference is that one can only measure or probe a particle by using a comparable quantity, a quantum of light. A quantum of light will disturb the position of a particle. So, one cannot accurately determine the position and velocity of a particle of light simultaneously. Continuous efforts to determine position and velocity mean that the more accurately the position is determined, the less accurate the velocity is determined. Quantum Mechanics had grave consequences for Determinism. If one cannot measure the present velocity and position of a particle, one cannot predict its future velocity and position. If one cannot measure the present state of the Universe, one cannot predict its state at any other time. In Quantum Mechanics, velocity and positon are a combination of both. Thus Quantum Mechanics does not predict a definite result of an observation, but a possible result, hence the Uncertainty Principle. From a Theological point of view, what is significant about Quantum Mechanics? The significance is implied in the fact that there are two great variables in the Universe – Matter and Intelligence. Since matter is so intelligently arranged, it is possible that it has an intelligent cause. Why cannot scientists apply Quantum Mechanics to the Universe and say it has an intelligent cause? And since the stability of the Universe is never questioned by the greatest scientists, why cannot scientists think the Universe is governed by an Intelligent and Immutable God? Those scientists who are so sure that the Universe is expanding and who express strong conviction in the accidental evolution of the Universe should reconsider the Theory of Quantum Mechanics.

The beauty of the Theory, from my point of view, is that we can apply it to everyday life. It means we must use our discretion in the daily circumstances of life. The Uncertainty Principle in Quantum Mechanics suggests that the velocity and the position of a particle of light cannot be independently measured. A discretionary approach is taken to combine both velocity and position. So in daily life, we should act within the limits set by Quantum Mechanics. What this means is that we should always give someone else the benefit of the doubt, but we should not go to an extreme to do it: for to do so would not be within the limit allowed by Quantum Mechanics. We still need to use our discretion when we give someone else the benefit of the doubt.

CHAPTER 4
The Mind of God :
As Revealed by a Schedule of Laws Descriptive of the Universe

Although we encounter many mysteries in trying to understand the Universe, the Universe is not mystical – it is a Universe of matter. Being so, at a certain point in its formation, it had to form according to the laws of science as we have already pointed out in the Theological Postulate. And as we have therein formulated the premise: Every phenomenon in nature, event, occurrence, or condition is the result of some other phenomenon, event, occurrence, or condition of nature and can thus be naturally explained. We want to consolidate that Premise with the Premise of this chapter that: *All the Laws of Science Were Developed in the Mind of God in Eternity Past and Began to Materialize the Moment God Gave the Command of Creation.*

The interplay between heat and cold, expansion and contraction, velocity of motion and the momentum from the weightlessness of outer space gave cause to the three forms of matter, solid, liquid, and gas. God had pictured the present state of the Universe in Eternity Past. An imaginary Universe became a natural reality. We perceive the following as some of those laws developed in the mind of God and became operative immediately after He created the conditions of matter. The present state of the Universe was not surprising to God. In the infinite mind of God, He foresaw the present state of the Universe in all its glory and beauty. He foreknew the laws of science developed in His mind would have guaranteed the stability and beauty of the Universe. But what were some of those laws? Some of those laws were The Law of Weightlessness, The Law of Gravity and Motion, The Law of the Elements, The Law of Variations, The Law of Distance and The Law of the Horizontal Position of Bodies in Space.

In the Twentieth Century, the highest level of civilization, our scientists have found by experience that conditions of weightlessness prevail in outer space. This state of weightlessness is critical to the motions of the stars and other bodies in space. Everything in our Universe is an effect of some natural cause. If the Universe is in a state of weightlessness, there must be a natural cause or explanation.

The miracle of Creation first occurred: the conditions of matter [space and heat] being created. We can now follow the steps in the formation process. The whole spatial Universe was hot, matter in the form of gas [heat]. It then cooled. In the cooling process, most of the matter formed into dark invisible matter beneath while condensation in the form of clouds and dust occurred above. As the formation process continued, the dark invisible matter took a more stable form. At least 75 percent of the dark invisible matter would become stable. The matter of clouds of dust would float above. Only about 25 percent of the hot gas contracted into clouds of dust. So more than 75 percent of the matter in the Universe is dark invisible matter. It is amazing that the amount of hot gas cooled into clouds of dust of far greater density than the dark invisible matter. But is it not equally amazing that when a cup of hot milk cools, less than one percent forms into cream? We can describe this early formative stage of clouds of dust and gas as the Cosmic Cream.

All evidences point to the fact that the Universe contains more invisible matter than visible matter. The weightlessness of outer space, the Microwave Radiation, the radiation of the stars, are evidence to the fact that there is more invisible matter than visible matter in the Universe. The average star converts 4,000,000 tons of matter into energy every second. In the realm of nature, many things do exist and do happen which are determined by their effects. For instance, the Earth rotates at a speed of 1,000 miles per hour and orbits around the Sun at a velocity of 66,000 miles per hour, yet we do not feel the motions of the Earth. This is indicative of the fact that the Universe is upheld by God's omnipotence.

It will become more evident that the Universe has more invisible matter than visible matter and that the invisible matter is the foundation of the Cosmic Structure, as this discussion continues. So it will be a question as to what percentage is invisible matter. Well for an illustration, let's look at the mass of the Sun to see how much volume of matter it contains and then convert this volume of matter into the form of gas [the equivalent of invisible matter]. The volume of a sphere is 4/3 x 3 1/7 x T3. The Sun's diameter is 860,000 miles. Its radius is 430,000 miles. Approximate volume of matter would be 430,000 x 430,000 x 430,000 x 4 cubic miles [4/3 x 3 1/7 x 430,000 x 430000 x 430,000]. We convert the volume of matter to liquid form by multiplying it by 1,000 [1,000 x 4/3 x 3 1/7 x 430,000 x 430,000 x 430,000]. Convert to the form of gas by multiplying the liquid form by 1,700 times [1 with about 30 zeros] [1,700 x 1,000 x 4/3 x 3 1/7 x 430,000 x 430,000 x 430,000].

The conversion of the Sun's mass into gas form [equivalent to invisible matter], would occupy an infinitesimal amount of space [like an atom in the infinity of space]. Furthermore, the depth of the volume of space occupied by the Sun's conversion into gas form would have to be measured in depth of light years to create the weightlessness in the Universe.

This volume of space when compared with the volume between Proxima Centauri (the nearest star to us) and the Sun is indeed infinitesimal. Proxima Centauri is approximately 4.5 light years away or about 26 trillion miles (26,000,000,000,000) from the Sun. The distances between many other stars are much farther. This means that if all bodies in space were converted to the form of gas (invisible matter), it would only occupy about 25 percent of the space in the Universe!

You know thousands of years ago, with a great smile, the CREATOR asked Job the question, "Where was thou when I laid the foundations of the Earth? Whereupon are the foundations thereof fastened?" (Job 38:4-6). Who would have thought that in time man would have discovered the most profound secret of the Universe? The wonder and the marvel of it all is that nature with all its laws hinges on the omnipotence of God's word.

The Law of Weightlessness

The invisible matter became the foundation of the Universe by causing the condition of weightlessness of outer space. It is like ships in the oceans. The volume of water occupied by a ship is heavier than the ship, so it floats on. On this principle numberless ships sail the oceans. Thus all bodies in space are in a state of weightlessness. This fact can be illustrated from our daily experience. Each time we look at the skies we see volumes of clouds moving in all directions. We know that this has been possible because of the air pressure of the Atmosphere. We do not see the Atmosphere but we know it is there and causes the clouds to move. We can also think of a beautiful fair day that reminds us of the beauty of nature and the pleasure of life.

But this is only a transient reminder because when evening *comes and the thunder begins to sound, the harmless elements* of the Atmosphere become so violent that most of the time the place where we sit or stand and the whole area around us begin to tremble. If we had not learned from previous experience, we would not have imagined that the elements of the Atmosphere would have produced such dreadful effects of thunder and lightning. So we can clearly see how the Dark Invisible Matter in the Universe can and has produced a state of weightlessness.

Specifics of the Law:

1 The dark invisible matter is far greater than all other forms of matter combined to the extent that all other forms of matter float in.

2. The weight and motions of a mass give that mass a systematic behavior in the weightlessness of outer space: it does not float aimlessly.

3 All small masses of matter within the area of a large mass are not subjected to weightlessness because of the gravitational attraction of the large mass. Weightlessness is directly related to distance. If they are removed far enough from the large mass, they will be subjected to condition of weightlessness.

4 When weightlessness applies, it applies absolutely. All objects of matter within a large mass in outer space are subjected to weightlessness. It is as though no door can be used to close out weightlessness. All objects of matter within a space ship have to be tied down to prevent them

from flying around.

  5 Any mass of matter in outer space will warp or carve space to the proportion of the size or volume of that mass of matter, because space is filled with dark invisible matter. With regards to the Solar System, not only the mass of the Sun has warped space, but also all nine planets. The sky we see above us is due to the fact that the Universe is filled with dark invisible matter and at some point in infinity had to show up, had to manifest itself. It has done so in the form of the sky. The sky above us is also due to the fact that stars have so warped space to the degree that it appears like an umbrella over the Earth. The sky is really the Cosmic Umbrella. Because the Universe is filled with dark invisible matter and because any mass of matter will warp space, the Universe will look the same from whatever location it is viewed.

The dark invisible matter not only causes condition of weightlessness, but is also important from the aspect of absorbing most of Cosmic Radiation (radiation of stars). In the process great clouds are formed. The dark invisible matter is mostly cold with temperature of a few degrees above Absolute Zero (524 F or 273 C below zero). The distance from a radiating body will determine the degree of the temperature. When Cosmic Radiation focuses on dark invisible matter, billions of tons of matter are lifted into the skies. Through the action of Gravity and Motion some of these clouds are later placed just in a position to absorb Cosmic Radiation to the degree that they become luminous. However, the dark invisible matter absorbs most of Cosmic Radiation. Another reason is the fact that the Universe is measureless. This answers the aged old question, "Why is space not so filled with radiation as to make the nocturnal sky look like a glowing surface?" There will remain many unanswered questions about the Universe.

The Laws of Gravity and Motion

The Laws of Gravity and Motion were intertwined with the formation of the Universe. They were discovered by Sir Isaac Newton, but not adequately explained. We do know that these laws were in the mind of God before He created the Universe. In Chapter 3 these laws were brought into focus and it was suggested that motions originated during the formative state of the Universe. That best explains the reality of Celestial Motions. Herein, the emphasis will be on the evidence of Gravity and on the HOW, rather than its definition.

The Earth itself offers the simplest and best evidence of Gravity. If an object is placed above the Earth within distance of feet to a mile, that object will fall right to the ground on its own, with the help of Gravity. However it is necessary to consider Gravity within a much larger context. The Gravity of a mass is determined by its size against the background of the weightlessness of outer space. The Universe is in a state of weightlessness. All masses are thus affected and the objects of a mass are controlled by its Gravity. Furthermore, the Gravity of other masses within a common neighborhood acts upon each other, depending on distances. The Solar System is a good example. Whereas we cannot work against Gravity, we can always work with these laws. The technology of the airplane has allowed it to work with Gravity, taking travelers to their destination all over the World. This technology has even taken man far beyond. For this to happen the airplane must be in good mechanical condition. In the event of a mechanical failure, it will immediately fall to the ground by the force of Gravity. Everything about the plane was designed to work with the Laws of Gravity and Motion. Its lift off, its control of air flow, and its speed were all designed to counterbalance Gravity.

The birds were created to work with the Laws of Gravity and Motion. Their wings and feathers were designed to give them the thrust and sustain their momentum in the Atmosphere. But in the event that the bird became disabled, it would fall speedily to the ground with the help of Gravity. We live and die by the laws of nature, so it is best if we become friends with these laws. This means the longer we will live or the sooner we will die. We have done a good job in learning to live with the laws of nature.

Already we have seen the importance of working with the laws Gravity and Motion. Living in Earth's environment is much easier to live with the Laws of Gravity and Motion. For one reason, we do not feel the effect of the Earth's motion —— it is just too fast for our sensibility: that is a good thing. Leaving the boundaries of the Earth, things get more difficult and the Laws of Gravity and Motion would have to be more carefully observed. We fly East and West, North and South without experiencing any noticeable difference. Getting from the Earth's boundaries to outer space, things get more difficult because one would first have to travel in the direction of the Earth's motion. To do otherwise would mean that the spaceship would be un-mercilessly thrown out of its orbit to the destination of **Doom's Day of No Return.**

At this time the Earth's motion would be felt with great furry and very little or nothing would be left of the spaceship. The Earth rotates at speed of 1,000 mph. The second thing one would have to consider is how to escape from this velocity because the Captain of the spaceship would not get very far if he attempted a speed of 23, 900 mph, or even 24,000 mph. There is a certain velocity he must attain. It is called the Escape Velocity. The Escape Velocity is defined as the minimum speed needed for an object to "break free" from the Gravitational Attraction of a massive body. Also, the Escape Velocity is the speed at which the sum of an object's kinetic and its gravitational potential energy is equal to zero. The Escape Velocity from the Earth is set at 25,020 mph. This means that at this velocity, the Earth has no Gravitational effect on the spaceship. Having said this, we can look more generally into this Universal Mechanism. Understanding it fully will enable us to answer other fundamental questions about the Universe. Questions were made to ask and each one demands a direct and complete answer. Each answer will inevitably lead to a greater understanding of the Creator.

Gravity in some way works like a piece of natural magnet. So to understand how a piece of natural magnet works is to understand how Gravity works. This will explain the 23 degree tilted angle of the Earth. Magnet has magnetism, that quality to attract metallic objects unto itself. The metallic object must have some properties of magnetism in order to be attracted but not enough to attract other metallic objects. The attraction of the magnet seems highly mysterious. But it must be obeying a definite law of nature. The area around the magnet is magnetized, a magnetic field surrounds it. A magnetic field also surrounds the metallic object. What then seems to happen is that when the magnet and the metallic object are brought within a certain proximity their respective magnetic fields interact. The magnet which has the stronger magnetic field pulls the metallic object unto itself.

The electromagnetism of matter extends outside the body of a mass. This means that a magnetic field surrounds every mass of matter. Some magnetic fields are transparent. The Earth's magnetic field is an example. It converges in the Ionosphere of the Atmosphere. Here the air becomes electrically charged particles. A magnetic field surrounds the Earth. This is important to radio transmission and to its gravitational connection to the Sun. But any mass of matter has a magnetic field independent of another mass.

But how does the principle of the attraction of magnet apply to or explain the 23 degree tilted angle of the Earth? If you suspend a piece of wire in a state of weightlessness and place a magnet close to one end, that is just close enough for it to be attracted, one end of the wire will be tilted towards the magnet while the other end will be tilted from it. The attraction of the magnet is focused at one end of the wire. If the wire and the magnet were placed parallel the one to the other, the wire would be attracted to the magnet without one end being tilted in either direction towards the magnet.

The Earth and the Sun and, indeed, all bodies in space are in a state of weightlessness. The gravity of the far greater mass of the Sun is focused at the North Pole of the Earth. This causes the Earth to be tilted to the proportion of the Sun's gravity focused at the North Pole. The Earth cannot be attracted unto the mass of the Sun. The velocity of its motions (1000 mph. rotational speed and 66000 mph. orbital speed) balances the attraction of the Sun's gravity. The Earth, therefore, maintains its tilted angle of 23 degree.

When Einstein introduced his famous Cosmological Constant, a force he felt was necessary to balance the attraction of Gravity; he explained that matter had a pressure point which prevented it from becoming infinitely dense. But he made the mistake of suggesting that it extended outward to balance the attraction of Gravity. He later withdrew this suggestion calling it the greatest blunder of this life. He explained that his Cosmological Constant was deuced from Maxwell's Theory that electromagnetism could cause singularities or infinite density of matter. If Einstein had just looked at the other side of the coin, as well as other scientists, they would have seen that the electromagnetism of matter extends outwardly, creating a magnet field for each mass of matter. I guess they might have thought it might have hastened the collapse of the Universe. Well, by the same token we know that the Universe could not have accidentally evolved; it would have collapsed a long time ago. The fact is that Gravity is a reality of nature. And in as much as it is a reality, must be naturally explained. The only natural explanation as I understand and as I can see is that the electromagnetism of matter extends outwardly, creating a local magnetic field. This magnetic field interacts with the Magnetic-field of the Universe.

Tilted position of wire when magnet is focused at one end of wire

Figure 4.1

The attraction between a magnet and a piece of wire in a state of weightlessness, explaining the 23 tilted angle of the Earth.

MAGNET

While one end is tilted inwardly the other end is tilted in the opposite direction. This also illustrates Newton's third law of Motion: Every action causes an equal and opposite reaction.

Figure 4.2

<u>Erect position of wire when magnet is focused at center of wire</u>

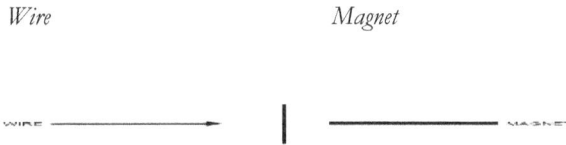

Wire   *Magnet*

WIRE ─────────────▶ | ──────────── MAGNET

To understand Gravity on the large scale is to understand it on the quantum level. And to say Gravity works over long distances without describing its nature and without describing a medium through which it works, is not really understanding it on the large scale. The Theo-Cosmos observations explain both levels. The Magnetic-field of the Universe means that space is filled with electromagnetic waves as a result of star radiation. Within the electromagnetic waves, there is a flow of Gravity at the quantum level. Between two particles there is also gravitational attraction. This extends to mass to mass gravitational attraction.

The Gravity between particles has been described as the weakest of the Four Elementary Forces. In a practical sense there is no difference between the Gravitational Force and the Electromagnetic Force because Gravity can only be understood as a magnetic attraction between two particles or two masses. However the significance of showing that Gravity is the weakest of the Four Forces is that if it were stronger the Universe would have already collapsed, the velocity of Motion would not be great enough to balance the attraction of Gravity.

So, how does Gravity extend from a particle to particle attraction to a mass to mass attraction? Through the electromagnetism of two masses which extends beyond them creating around them two magnetic fields and through the electromagnetic waves from outer space two masses are attracted the one to the other.

Let us illustrate this by using the Sun and the nearest star, Proxima

Figure 4.3

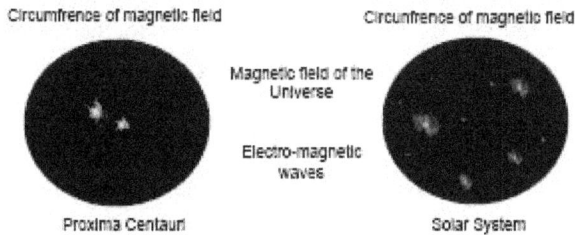

Circumfrence of magnetic field   Circumfrence of magnetic field

Magnetic field of the Universe

Electro-magnetic waves

Proxima Centauri   Solar System

Centauri. Let's say the magnetic field of the Sun extends to the circumference of the Solar system and the magnetic field of Proxima Centauri extends around it in a similar circumference. The space between the two magnetic fields is filled with electromagnetic waves from outer space. The gravitational attraction between these two bodies works through their respective magnetic fields and through the electromagnetic waves from outer space, the Magnetic field of the Universe. The Magnetic fields of two bodies interact with the Magnetic-field of the Universe to effect gravitational connection. Diagram shows the circumference of magnetic-field of Proxima Centauri and the circumference of magnetic-field of the Solar System.

The electromagnetism of matter at the quantum level is not strong enough to make the atom infinitely dense. The principle being the same as the reason – electromagnetism extends outwardly. It follows naturally that if the electromagnetism at the quantum level does not cause infinite density, on the large scale level it cannot cause the Universe to collapse. Gravity works over long distances between two bodies great or small through the Magnetic-field of the Universe. Gravity can only be described as electromagnetic attraction. The Magnetic fields of two bodies interact with the Magnetic-field of the Universe to effect gravitational connection. The question of instantaneous Gravity was once asked of Einstein, to which he replied that Gravity was not like any other force. Einstein was wrong because though Gravity would have instant effect on an object at its instant removal, there would have to be a reactionary time for it to have complete effect on that object, depending on the size of the object. All masses do not have the same degree of electromagnetism. Bodies like the stars are 100% electromagnetic. This is one reason that their distances are light years apart. The great distances between bodies in space have helped to balance the attraction between these bodies. The velocity of motion has also helped to balance the attraction of Gravity. Then there are the great Centrifugal and Centripetal Forces of the Universe which have also helped to balance the attraction of Gravity.

Everything in the Universe is interrelated. Nothing works within a vacuum. The invisible matter is the greater amount of matter in the Universe. It is equally true that it must play the greater part in the system of the Universe. We have clearly seen that the invisible matter combined with the weightlessness in outer space to form a single object called, "The Foundation of the Universe." With distances stretching from billions of miles to light years, great velocity of motions, and with the great Centrifugal and Centripetal Forces of the Universe, we are guaranteed a free Universe, a Universe free from gravitational collapse.

Before we consider the Laws of the Elements, let's note the Theo-Cosmos Observations of the Laws of Gravity and Motions.

1. The Electromagnetism of Matter Extends Outwardly of a Mass of Matter, Creating a Magnetic-field.

2. Gravity Works Through the Magnetic-field of the Universe in Conjunction with the Magnetic-fields of Two Masses.

3. When the Mass of the Attracting Body Is Round, Its Gravitational Center is Within Equal Distance at All Points: Its Magnetic-field Is Equally Strong. An Orbiting Body Will Orbit in a Circle.

4. When the Mass of the Attracting Body Is Elongated (Egg shaped), the Orbiting Body Will Orbit in an Elliptical Circle.

5. When an Orbiting Body Gets Closer to the Center of Gravity of the Attracting Body, Its Distance Decreases and Increases While It Orbits Away from the Center of Gravity of the Attracting Body.

6. A Body Will Orbit the Closest Center of Gravity. A Star Will Orbit the Nearest Star or Constellation of Stars. The Size of a Mass in a Group of Masses Will Determine the Center of Gravity.

7. When the Orbiting Body Is in Parallel Alignment with the Center of Gravity of the Attracting Body, the Orbiting Body Will Orbit in a Regular Position.

8. When the North Pole or the South Pole of the Orbiting Body Is Aligned with the Center of Gravity of the Attracting Body, the Orbiting Body Will Maintain a Tilted Position, Like the Earth and the Sun.

The Law of the Elements

The law of the elements took effect immediately after the conditions of matter [space and heat] were created. The creation of the condition of matter means that the whole spatial Universe was intensely hot, matter in the form of gas. There was only one element. If there were more, they would have been unified at this temperature. From here we can deduce the Law of the Element, how more than 92 elements found on Earth were formed from 1 to 3 original elements. Hydrogen is the lightest and most abundant element in nature. It is the most abundant element found in stars except for the Neutron Star and the White Dwarf Star, which are the minority of stars.

Water covers about 75% of the Earth's surface from a few feet to 7 miles in depth. Two-thirds of water is made of hydrogen atoms. So the liquid part of Earth is called the Hydrosphere. Protein is very critical to organic life. Amino Acids are the building blocks of protein. The basic structure of Amino Acids includes, among other atoms, a group of Hydrogen Atoms. Most of the human body is made of water. This means that in the Universe and in the human body Hydrogen atoms are the most abundant atoms.

The scientific evidence seems to indicate that Hydrogen was the original element from which all other elements originated. But there is no scientific explanation as to why the original source matter should take various forms such as stars and planets. If stars were the only things formed from the embryonic state of matter, science could explain. But how can science explain how the embryonic state of matter produced a radiating mass and a mass of solid rock? The reality of this variation of two different kinds can only be understood if it is considered to be by Divine purpose. So what the Law of the Element will explain is the formation of the elements but not the different forms of matter (a radiating mass and a mass of solid rock).

In his book, Galaxies, Nuclei, Quasars, Fred Hoyle has shown how at different degrees of temperature most of the elements found on

Earth are produced in the stars. The Sun sustains a maximum temperature of 20,000,000 million degrees Centigrade. At this temperature Hydrogen is the most abundant element in its atmosphere and in its nuclear fuel. At this temperature Hydrogen fuses to produce Helium. This releases great amount of energy. At temperature of 100,000,000 million degrees, Helium can fuse to produce carbon, oxygen, and neon. Following Helium exhaustion, at a higher temperature carbon and oxygen can produce such elements as sodium, magnesium, aluminum, silicon, and sulfur. There is only one more stage before the end in the stages of the nuclear fuel provided a temperature of 3 billion degrees is attained. At this temperature the "iron group elements" are found such as Chromium, manganese, iron, cobalt, and nickel. Very few stars may sustain a temperature of 3 billion degrees. The Regular Sequence stars like the Sun have a temperature of 20,000,000 degrees Centigrade.

Ninety two elements are found in the Earth's crust which has a depth of 20 miles and a temperature at the bottom of 100 degrees Centigrade. The dominant elements found in the Sun's atmosphere (gaseous part) are Hydrogen and Helium. The Sun's atmosphere is more than a million degrees Centigrade. What is becoming clear to us is that temperature is a factor in the difference in number between the elements in the Earth's crust and those in the Sun's atmosphere. Therefore to understand the Law of the Elements we must begin with one element and with a temperature that will guarantee one element of matter. We here state the specifics of the Law of Elements thus:

1 At absolute temperature all elements are indistinguishable, one element. This temperature can convert all metallic elements into the form of gas. When thus converted, they will be recognized as one element. This law presupposes that the embryonic state of the Universe was a state of gas at absolute temperature.

2 At various levels of temperatures and volumes of matter,

formed into various elements and combinations of elements. However, because of the temperature of the Sun, it has fewer elements than the Earth. In the petroleum refinery, the oil is heated into the form of gas at about 700 F. The gas is then cooled at various levels of temperature. This produces different chemical products.

3 The chemical reactions of some combinations produced other elements. In light of the Laws of the Elements, we have no choice except to reject the scientific deduction of some scientists that an earlier generation of stars was necessary to create the heavier elements of the present generations of stars.

The Laws of Variations

The Laws of Variations we shall herein propose do not explain our Universe of amazing varieties. Simply, science cannot explain why matter should be in different forms or kinds. The only explanation to the different forms of matter is a Theological one. Different forms of matter were by Divine purpose. However, these laws determine the variations in each different form of matter. In the formation of the elements, the stars and the planets, the laws of variations apply. The human race is no exempt nor is any other thing in nature in the microcosmic world and the macrocosmic.

We present the specifics of the Law of Variations thus:

1. The nature and volume of matter determine the variations in that particular form of matter. The greater the volume of matter, the more and diverse are the variations. But to a great extent, this depends on its nature. A chalk stone when falling from a certain height onto a hard surface will have broken into several pieces and each will have varied in size and shape.

If a larger chalk stone is falling from a higher height, it will have broken into many more pieces, each piece varying much more from the other. If that larger chalk stone was more dense or harder than the smaller one, it might have broken into the same number of pieces, or even fewer numbers of pieces, depending on its density or hardness.

Even in the Cosmic Background Temperature (the temperature of outer space of 454 F below zero), there are variations due to the immensity of the Universe. It was no surprise to me when scientists reported in April, 1992 that they found variations in the Cosmic Background Temperature. What was a surprise to me was why they had not thought of that before. What was rather astonishing was their interpretation of these variations that they were "relics of the Big Bang". There are variations in all of nature because there are Laws of Variations.

2. The process of formation of a particular form of matter will determine the variations of that form of matter. All kinds of coal are made from the original matter of dead plants and animal matter. But different stages in the process produce coal variations. Peat is mainly a mixture of carbon and water with other minerals. Brown coal is a further process of Peat. Soft Coal is a further process of Brown Coal. When all the water is processed out of coal, the final result is hard carbon.

3. The life span of a particular kind of matter will determine variations. (a) At the beginning of the life span of a particular kind of matter variations are less diverse. (b) At the middle of the life span of a particular kind of matter variations are more diverse. (c) At the ending of the life span of a particular kind of matter variations are most diverse.

The Law of Horizontal Position of the Universe

The immensity of the Universe puts all bodies in outer space in an horizontal position. The whole Universe lies in a horizontal position. Saying this, we have in mind, according to some estimate of scientists, at least 200 billion galaxies each having at least 100 billion stars. The expression, "the Table of the Universe and the One Who Sits There", best enables us to understand the horizontal position of celestial bodies. These bodies appear vertically above us. This is not a deception of our imagination but is simply the fact of viewing the Universe in a finite way. But when viewed in an infinite way, the Universe lies in a horizontal position. The Creator sees the Universe in these two positions.

Imagine that we sit around a large table and we use the table cloth as a map of all the galaxies. We spread the table cloth over the table. We look down on the table and we see all the galaxies of stars lying in a horizontal position. But this does not only explain the infinite reality as to the position of the galaxies in relation one to the other, but also the reality of viewing the heavenly bodies in a finite way: the vertical position they appear to us here on Earth. God looks at the Universe in both ways. He sits around the Table of the Universe and sees it perfectly and completely: the way we human beings see it from the surface of the Earth and the way it is seen in infinite reality – the horizontal position, the Galactic Plane.

In the Moon's orbit around the Earth, it sometimes appears directly above the Earth. But they both lie in a horizontal position. Because of their sizes and distance apart, the one appears to be directly above the other. It all depends on what location from which they are viewed. If you were on the Moon looking at the Earth, it would appear at times directly above the Moon. When the astronauts landed on the Moon, the Earth appeared directly above the Moon, reflecting the light of the Sun. This is proof that they both lie in a horizontal position and is to be further proof that all other bodies in outer space lie in a horizontal position. If there were a rocket with the capability of landing on anybody in outer space and if it were directed in a vertical position, it would have traveled for all eternity without landing on anybody in outer space. Again, this argues the fact that all bodies in outer space lie in a horizontal position.

The Law of Distance

Distance is simply the space between two things. It is one of those overlooked laws of nature. Yet it pervades the whole realm of nature. To broaden our definition, we shall consider Distance in three dimensions: Microcosmic Distance, Solar Distance, and Macro-cosmic Distance. Microcosmic Distance is the distance between particles. The electronic construction of the atom is a good example. The atom has been redefined to be consisted of a core of protons and neutrons with a number of orbiting electrons. Furthermore, protons, neutrons, and electrons are no longer considered the most basic particles of matter.

The protons and neutrons are made of more basic particles called, Quarks. Each is made of three quarks. The electron is found with an anti-particle called, Positron. The distance between particles means that the atom is not absolutely dense.

Solar Distance is distance of miles and fractions of a mile (inch, foot yard, etc.). This is the distance with which we are most comfortable. We use this standard within the Solar System, hence the name. Outside the Solar System, we must consider another dimension – the Macro-cosmic Distance. The Macrocosmic Distance is distance of light years. Since this dimension of distance applies to Astronomy, we can also call it Astronomical Distance. It is used of stars and galaxies. A light year is the distance light travels at the speed of 186,000 miles per second in one year. A light year is approximately 6 trillion miles.

The stars in the Universe are equally spread and the galaxies are evenly spaced. This does not mean that all the stars are the same distance apart, nor the galaxies. It means that they are evenly spread in a general sense. The distance between two stars or bodies in space guarantees that they do not collide. Each time we look at the heavens we are looking at the laws of Distance. Each time we consider the electronic construction of the atom, we are considering the Law of Distance.

Specifics of the Laws of Distance:

1 A Particle cannot be too close to the other.

2 A Particle cannot be too far from the other.

3 A particle must be at the effective distance. A star cannot be too close to the other. A star cannot be too far from the other.

A star must be at the effective distance. These three overlooked laws of Distance underlie the whole realm of nature. By effective Distance we mean the distance that a particle or star must occupy to be an effective part in the system of the Universe. With respect to the radiant bodies, their magnetic attraction explains their distance of light years one from the other.

CHAPTER 5
The Structure And Forms of Matter

To have an understanding of the Universe is to have a knowledge of the structure and forms of matter. Already the size of the Universe has been emphasized. The cause of weightlessness in outer space has already been identified. Newton's laws of Gravity and Motion have already been explained with some specific observations to give a better understanding. In this chapter we shall look at the structure and some solid forms of matter. Matter exists in three basic forms: solid, liquid, and gas. The solid forms of matter that will be herein discussed are the Solar System, the stars, Black Holes.

The Atom

It is important to first define matter within the Theo-Cosmos context. It has been already pointed out that critical to the formation of the Universe were the conditions of matter: space and heat. These conditions represent the Miraculous Aspect of the formation of the Universe. From these conditions all the 92 commonly known elements of matter were formed. Until the Eighteenth Century the atom was considered the smallest unit of matter. The Nineteenth Century witnessed a change of definition. By experiments, under microscopic examinations, scientists began to establish that the atom was made of several particles. They have further defined these particles into three different categories, namely the protons, neutrons, and electrons.

The protons are positively charged particles. The neutrons are uncharged particles. The electrons are negatively charged particles. The protons and the neutrons form the nucleus of the atom. The interaction between the neutrons and the protons creates a force which causes the electrons to orbit around it. This construction of the atom was founded on the theory that like particles repel each other. The electrons left alone would repel each other, so as the protons. The neutrons the uncharged particles exert a force on the protons, overcoming the repulsion among them and thus unite with them. The positively charged protons of the nucleus attract the negatively charged electrons. This attraction of the protons keeps the electrons orbiting the nucleus. Although the protons are equal in number to the electrons, a proton is 1800 times the mass of the electron. The neutron has a greater mass than the protons.

Although the Theory of the Atom was well established during the Nineteenth Century, its inner structure was discovered by its behavior under microscopic examination. This was not the result of a single experiment but a number of experiments over a period of time. By 1911 the physicist Sir Ernest Rutherford explained that the atom was made of a nucleus of positively charged protons with a number of orbiting electrons. By 1932 it was further discovered that the nucleus has other particles that were not positively charged but have the identical mass of the proton.

In the same year it was further discovered that the electron has an anti-particle, the positron. The protons, neutrons, electrons, and the positrons were then considered the ultimate particles of matter until in the early 1960's when it was further discovered that the neutron and the proton were made of tiny particles called, "quarks". A proton and a neutron is each made of three "quarks".

Elementary Forces

The structure of matter can be further understood by understanding its elementary forces. To fully understand the structure of matter is to understand how matter can be held together in all its forms eternally, or why matter is indestructible. The elementary forces are the forces arising from the interactions between the particles of matter. In such interactions a particle emits a force carrying particle which is absorbed by the other particle. This causes a change in the position and velocity of the particles involved in the interaction. The repulsive force between two electrons is caused by an interaction between their photons (quantum of light), in which case one electron emits a photon and the other electron absorbs it.

Scientists recognize four forces among the elementary particles and classify them according to their respective strengths. They are the Gravitational Force, the Electromagnetic Force, the Weak Nuclear Force, and the Strong Nuclear Force. The Gravitational Force is recognized as the weakest. Though Gravity is universal and inescapable, acting on masses of phenomenal sizes, when considered on a particle to particle relationship is the weakest. The Electromagnetic Force is the second weakest force but is millions of times stronger than the Gravitational Force. The reason for its superior strength is obvious. It is basically resulted from the interaction between the positively charge particles (protons) and the negatively charged particles (electrons). The electromagnetic attraction is so strong that it causes the electrons to orbit the nucleus.

The Weak Nuclear Force is much stronger than the Gravitational and the Electromagnetic Forces. When it is overcome or weakened radio-activity is released. The Weak Nuclear force is generally overcome when the energies of the particles exceed a certain limit. It operates at the level of the seam of the nucleus. The Strong Nuclear Force is the highest in the chain of command of the forces, and of such holds all particles of the atom together. At normal energy level it is at maximum strength. At high enough energy it gets weaker. When the Electromagnetic and the Weak Nuclear Forces exceed their respective energy levels, the Strong Nuclear Force begins to lose strength. If its energy limit is exceeded it can no longer hold the particles of the atom together. The result is inevitable – nuclear energy is released.

The strength of the Nuclear Force depends on the particular kind of atom. The Uranium 235 is held by a very strong force. Scientists had spent many years of experiments before they were able to overcome this force. When they had overcome it by splitting the atom, great amount of energy was released. The process of splitting the atom is called Nuclear Fission.

Before World War 2, Sir Ernest Rutherford and Sir James Chadwick attempted to split the atom but they were only successful in chipping it. Two German scientists, Otto Hahn and Fritz Stresemann were the first to succeed to split the atom. In 1939 they bombarded Uranium 235 with high speed neutrons and thus split the atom, releasing enormous amount of energy. By 1945 the technology of Nuclear Fission was used to produce the first atomic bomb. The World will never forget its disastrous consequences. Two hundred and fifty thousand people were instantly killed at Hiroshima, the target of the first atomic bomb. The World had never seen anything like it before.

Molecules, Elements, and Compounds

There are more than 92 elements found in nature, or more than 92 different kinds of atoms. Matter made of one kind of atoms is called an element. Matter made of two or more kinds of atoms is called a compound. Atoms differ in size and strength. The atomic number (the number of protons in the atom) and the atomic mass (the number of protons and neutrons) determine the size and strength of an atom. Each of the 92 atoms differs in atomic number and mass. The simplest atom is the Hydrogen atom. It has one proton and one electron. It is also the most abundant element in nature. One of the most complexed atoms is the Uranium U-238 which is used to make bombs. In atoms like the Uranium Atom the electrons are arranged into groups called shells. Different numbers of electrons are in each shell. The shell with the smallest number of electrons orbits closest to the nucleus of the atom.

Atoms are grouped together in groups of two and more. A group of atoms is called a molecule of matter. A molecule is, therefore, a combination of two or more atoms held together by a strong force called, Bond. This strong Bond is caused by the transfer of an electron or electrons from one or more atoms of that particular molecule of matter.

Gaseous atoms always group into molecules of two and three atoms. Different kinds of molecules form compounds. There are many kinds of compounds depending on the chemical combinations of atoms. The Hydrogen Atom is the lightest and simplest atom, and can combine with many different kinds of atoms. Two Hydrogen Atoms combine with one Oxygen Atom to form a molecule of water. A molecule of sugar is made of 22 Hydrogen Atoms, 12 Carbon Atoms, and 11 Oxygen Atoms. Some atoms will not combine with other atoms to form compounds. These atoms are more stable than others.

Radioactivity

A knowledge of the radioactivity of some elements will enable us to better understand the structure of matter. An element is considered radioactive if it produces a steady flow of radiant energy. Radioactive materials break down or decay at the level of the nucleus to form a more stable form of matter. The period of time for an element to form a more stable form of matter is called the half-life of that element. One gram of radium (1/32 of an ounce) will lose half its mass in about 1700 years and settle down to a more stable form. One gram of Uranium, U 238 will lose some of its mass in 5 billion years and settle down to the more stable form of lead. In the radioactive process matter is converted into energy and ejected at incredible speed.

In their book, Natural Science, Philip Gearing and Craig Conrad explain that all naturally occurring elements with an atomic number greater than 83 are radioactive. They explain three levels of radioactivity: Alpha Decay, Beta Decay, and Gamma Decay. The Gamma Decay emits very strong electromagnetic waves which are most dangerous of the three. What happens after radioactivity occurs? Only a small percentage of its protons is lost in the process. In the case of Uranium, U-238 with an atomic number of 92 before radioactivity occurs, now has an atomic number of 82 and an atomic mass or weight of 200.61, its final form of lead. What really happens after radioactivity occurs is that matter will last forever; matter is indestructible. It is of stupendous importance to here note that science concurs with theology on the indestructibility of matter. However, it must be pointed out that anything created can be destroyed. Naturally matter can last forever. But it can be destroyed if the CREATOR so chooses.

The Earth

The Earth

The Earth has the three basic forms of matter: solid, liquid, and gas. These three forms are known to all: those who can see as well as those who are blind. Before a blind man leaves home to church, he takes a shower: the liquid part. He then walks to the church next door: the solid part. While walking to church, he feels the gentle breeze: the gaseous part.

First, let us look at the Earth's gaseous part, the Atmosphere. The Atmosphere is the whole blanket of air which surrounds the Earth, extending approximately 400 miles above. It is made up of different gases and layers of air. The two most abundant gases are nitrogen (78 percent) and oxygen (21 percent). Only about one percent of the Atmosphere is made of other gases.

There are four layers of air. Each has a different volume of air. These layers of air are the Troposphere, the Stratosphere, the Ionosphere, and the Exosphere. The Troposphere has the greatest volume of air and extends about 10 miles above the Earth. To us human beings this is the most important layer. It is critical to the existence of birds and to air transportation.

The Stratosphere begins above the Troposphere and extends about 50 miles above. Its special peculiarity is that it has a gas called the ozone which converts the energy from the Sun into heat. The Stratosphere also shields us from the ultraviolet rays of the Sun. The Ionosphere Layer lies between the Stratosphere and the Exosphere. It begins above the Stratosphere and extends several hundred miles above. In it lies the Earth's magnetic field. Here the air molecules become ionized electrically charged, forming an electrical blanket around the Earth. The Ionosphere facilitates radio transmission. Without it there could be no radio transmission.

The Exosphere Layer begins above the Ionosphere and extends

to four hundred miles above the Earth. It has very little air molecules; and because of this, the temperature is very hot. There is no special significance about this layer. With each layer of air, the volume decreases. The Troposphere the first layer has approximately 80 percent of the volume of air surrounding the Earth. This layer is the most dense, due to the Earth's gravity. But as the Atmosphere gets farther and farther away from the Earth's surface its density gets less and less.

The Hydrosphere: the Liquid part of the Earth

About 75 percent of the Earth's surface is covered with water. Water is a compound of Hydrogen and Oxygen Atoms. Two thirds of it is made of Hydrogen Atoms, hence the name, Hydrosphere. The Earth's surface covered by the water is called the ocean's basin. But it is not as smooth or as flat as the bath tub or kitchen sink. It is made of great valleys, plains, mountains, and deep trenches. It is just like the surface of the dry land. Therefore the depth of water will vary greatly. It varies from a few feet to seven miles. Then some of the deepest parts have trenches several miles deep.

The Hydrosphere is important for three main reasons. First, it supports the greater percentage of animal life. More animals live in oceans than on land. Some of these animals provide a vital source of food for us. Second, the oceans are the source of the water cycle, rain falls. Billions of tons of water evaporate each day into the Atmosphere. After forming into rain clouds it falls again into the oceans and on land. Third, the oceans help to maintain a balanced temperature. It absorbs much of the Sun's energy to the proportion of the surface it covers. Otherwise, the temperature would have been much hotter. A portion of Sun's energy captured by the oceans is released daily into the Atmosphere. The Atmosphere spreads that portion of energy equally, causing a more balanced temperature.

The proportions of life conditions, the Atmosphere and the Hydrosphere are just the correct proportions. Without these proportions, life on Earth would not be possible.

The Lithosphere: the Solid Part of the Earth

With the exception of the Earth's crust of a depth of 20 miles, the rest of its structure is made of four layers of rocks. Each of these layers differs in volume, density, and temperature. The Earth's crust is believed to be once itself a solid layer of rock. Geologists suggest that the Earth was formed from hot liquid rock. This volume of liquid rock cooled, condensed, and formed into different layers. The crust was later processed into what we know as soil chiefly made of rock particles. Small stones are also very common. It has a temperature of 1-1000 centigrade and contains about 92 elements.

The first layer of rock is called the Outer Mantle. It is just beneath the Crust. It has a depth of 600 miles and a temperature of about 1500 Centigrade. Second, is the Inner Mantle with a depth of 1,300 miles and a temperature of about 2000-3000 Centigrade. Third, is the Outer Core with a depth of 1240 miles and a temperature of 3000-5000 degree Centigrade. Fourth, is the Inter Core with a temperature of about 6000 degrees Centigrade and a depth of 800 miles. From the Crust to the center of the Inner Core is approximately 4000 miles.

The solid structure completes the picture of the Earth. The picture before us is an Earth with a diameter of 8000 miles, a circumference of 24,900 miles, and an area of 196,950,000 miles. It is covered with about 3,000,000,000 cubic miles of water, and surrounded with an Atmosphere of a depth of 400 miles. We can now reflect on the scientific description of the development of the Earth given by the geologists.

When we do, we shall see limitations in their creativity and imagination. It is clear for the very blind to see that a cooling Earth could create an Atmosphere and that hot Atmosphere could condense into the form of rain. But the geologists have not explained why only a part of the Earth was covered with water. The Bible explains the whole earth was covered with water and God gathered the water together into the oceans and the dry land appeared.

We can also see how the solid surface of rock could be broken up into fine particles or sand by the water that had covered the surface of the Earth. And we can also see why the soil is composed chiefly of sand. But someone may question this scientific deduction and say God could have covered the Earth with water without rain from this condensation. Such persons may want to remind us of the Bible Story of 2nd Kings, Chapter 3 in which God miraculously filled with water the ditches dug by the Kings of Israel, Judah, and Edom when they united in war against the King of Moab. After these ditches had been dug at the command of God, He miraculously filled them with water, yet there was no rain.

I personally believe in the miraculous power of God, His omnipotence. And I truly believe this miracle took place. But to understand this book is to understand the two great contradistinctions in the formation of the Universe: the Miraculous Aspect and the Scientific Aspect. Following the Miraculous Aspect, the Universe had to form according to the laws of science. We can assume the Scientific Aspect involved most of the time. We recognize that God might have put the laws of the Universe at work and left them working. In such case these laws would respond to God's command with a sense of urgency. We also recognize there were times when God directly intervened and supervised parts of the formation of the Universe. We believe that the Solar System was one such part and one such time. Yet we believe that condensation in an hot atmosphere may have caused the Earth to be covered with water, rather than it being covered in a mysterious manner.

We must remember that God was not establishing a realm of mysteries; He was establishing a realm of nature.

The Solar System

It is believed that the whole Universe was once in a state of clouds of dust and gas according to the scientific information in Job 38. The solar system was once in that form. God used the laws of Gravity and Motion to form the Solar system as well as all other bodies in the Universe. If all bodies in the Universe were thus formed, they must have taken a round shape, which they did. With the force of Gravity pulling to its center clouds of dust and gas from all around, a volume of matter would eventually assume a spherical shape.

If this picture of the Solar System is correct, then the scientific description by the geologists is not completely accurate. They say that the Earth's crust was basically formed by volcanic activities spilling on additional hot liquid rock. Of course, they explain other factors in the formation of the Earth's surface such as weathering and the forces of glaciers. We can without hesitation concede the legitimacy of these forces as factors in the present state of the Earth's surface. Equally, without hesitation, we must point out that although these factors help to explain the formation of the Earth's surface, in reality, they do not explain the basic spherical shape of the Earth.

The spherical shape of bodies in space and their motions suggest that they were all formed by the same process and of such each would naturally contain most of the elements found on earth. The Sun shows an abundance of the elements, Hydrogen and Helium. This is due to the Sun's purpose and its temperature. It still shows other elements, some of which are detected by the Spectroscope.

Why are the diversities among the solar System? The answer is evident and that is why it has already been said and that is what is meant that God many times directly intervened and supervised parts of the formation of the Universe after He had set laws into motion.

The Sun

The Sun is considered an average star. It contains more than 98 percent of the matter in the Solar System. It has a diameter of 860,000 miles. It is in a gaseous state with varying temperatures at different parts. The shining disk is called the Photosphere which has a temperature of 6000 C. A layer of flaming gas called the Chromosphere, about 10,000 miles thick, lies above the Photosphere. Lying above the Chromosphere is another layer of gas called the Corona. It is about a million miles thick. Its temperature is about 1,000,000 C. The temperature in the Sun's core is estimated to be about 20,000,000 C. We can expect that the core is very dense because of the tremendous pressure exerted on it. By the process of Nuclear Fusion, the Sun converts some 4,000,000 tons matter into energy every second of time.

The Photosphere reveals a number of chemical elements. It contains 90 percent Hydrogen and 8 percent Helium and small amounts of other elements. The Corona has revealed such elements as iron, and calcium. The Sun makes one rotation in 28 days. Some parts may vary from this time period.

The Planets of the Solar System: Their Diameter, Temperature, Rotation, Orbit of the Sun, and Distances from the Sun

Mercury's diameter 3,100, 622 F, 59 days rotation, 88 days orbit of the Sun, 36,000,000 distance from the Sun. Venus's D, 7,900; 800 F, 243 rotation days, 225 days obit of the Sun, 67,000,000 miles from the Sun. Earth's D, 8,000 75–100 F, 24 hrs. rotation, 365 ¼ days orbit of the Sun, 93,000,000 miles from the Sun. Mars's D, 4,000 80 F, 241/2 hrs. rotation, 687 days orbit of the Sun, 142,000,000 miles from the Sun. Jupiter's D, 88,600, 10 hrs. rotation, 12 yrs. orbit of the Sun, 483,000,000 distance from the Sun. Saturn's D, 74,000, 10 hrs. rotation, 29 1/2 yrs. orbit of the Sun, 886,000,000 miles from the Sun. Uranus's D, 32,000, 240F, 11 hrs. rotation, 84 yrs. orbit of the Sun, 1,782,000,000 distance from the Sun. Neptune's D, 31,000, 280 F, 16 hrs. rotation, 165 yrs. orbit of the Sun, 2,793,000,000 miles distance from the Sun. Pluto's diameter and rotation unknown; 248 years orbit of the Sun; 4,000,000,000 distance from the Sun.

Jupiter, Saturn, Uranus, and Neptune are gaseous planets. Notice their speed of rotation. They are said to be made Hydrogen and Helium gases.

Because Mercury is the nearest planet to the Sun, it was thought to be the hottest. Venus is now considered the hottest. This is due to its thicker atmosphere which absorbs more of the Sun's energy. The four gaseous planets are made mainly of gases such as ammonia helium and methane, instead of rock. Six of these planets have moons. Earth has one: Mars, two; Neptune, two; Jupiter, twelve; Saturn, nine; and Uranus, five; making a total of thirty-one. Mars is smaller than the Earth, yet it has two Moons.

Comets and Asteroids

A comet is made of frozen particles and gases. The average comet has a diameter of one mile. When it orbits near the sun, some of its frozen matter is melted. When it moves away from the Sun, it begins to freeze again.

An asteroid is a small island made of rock and metal, orbiting the Sun. Asteroids vary in sizes. The largest has a diameter of 480 miles, while others have diameter of a mile and less. Most asteroids orbit the

Sun in the space between mars and Jupiter.

The Stars

A great amount of the matter in the Universe is in the form of stars. There are four main categories of stars: Regular Sequence Stars, Red Giants, White Dwarfs, and the Neutron Stars. The Regular Sequence Stars are about the size of the Sun. Most of the stars are Regular Sequence Stars.

They produce their radiation on the principle of Nuclear Fusion, converting matter into energy. The Nuclear Fusion of four Hydrogen Atoms produce one Helium Atom. In this process great amount of energy is released. The Red Giants are the largest stars, but by comparison are very few. The Red Giant, Betelgeuse is said to be millions of times the size of the Sun. They like the Regular Sequence stars produce energy on the principle of Nuclear Fusion, converting countless tons of matter into energy. This is what scientists know about the stars. But what do they know about the more than 1000 classes of stars from Revealed Science? What guarantee do scientists have that the White Dwarf and the Neutron stars were not thus created? Certainly, it reveals the limitation of science to answer the basic questions about the Universe.

The White Dwarfs are about the size of the Earth, very dense and hot. The average density is about 100,000 times more dense than water. A small container of this matter would weigh many tons more than a similar container of Earth's matter. Scientists believe that the White Dwarfs were once Red giants or Regular Sequence Stars, which after exhausting their nuclear fuel, have settled down to the stable form of White Dwarf. They explain that a White Dwarf radiates light on the principle of the repulsion of its electrons against its Gravity: repulsion balances the attraction of Gravity. When a star contracted to the size of the earth, its electrons cannot be further compressed. The electrons begin to exert pressure upward against the force of Gravity pressing inward.

The Neutron Stars

These are very special kinds of stars. Scientists discovered them because of their peculiar radiation. The Neuron star was discovered in 1968. It is the smallest and densest of all stars. They have a diameter of 10-15 miles. But they have an amazing density, gravitational force, and speed of rotation. A neutron star may pack a mass equal to that of the Sun. It makes about 30 rotations per second. It functions on the principle of the repulsion of its neutrons against its Gravity. Electrons are said to have been forced into its protons causing it to shrink to its incredible size and to continue to glow.

The American scientist J. Robert Oppenheimer, who predicted in 1939 its discovery, suggested that its gravitational force would be so strong that for an object to escape it, it would have to be traveling half the speed of light (93,000 miles per second). Oppenheimer also suggested that further contraction of the neutron star could increase its gravitational attraction to the degree that light would not be able to escape. How a star continues to glow on the principle of the repulsion of its electrons against Gravity (in the case of the White Dwarf); and the repulsion of neutrons against Gravity (in the case of the neutron star) is indeed an extraordinary phenomenon. If the White Dwarf and the Neutron stars were not thus originally formed, then God may have intervened in the course of nature to preserve His Universe.

The Pulsating Stars

Astronomers have discovered stars which contract and expand. These are called pulsating stars. The pulsating of a star causes changes in its brightness and also in its temperature. When it contracts it is not as bright as before but gets much hotter. When it expands, it gets much brighter and is less hot.

From the pulsating of a star one can deduce that the inner structure of stars, and in particular that of a pulsating star, is not equally and absolutely dense. Since all stars were formed from clouds of dust and gas and by the same process, one can expect to find great cavities in the inner structure, particularly the pulsating stars. This is natural to expect of large volumes of clouds of dust and gas contracting into stars. Even in the solid structure of the Earth there are many cavities great and small. Some of them are in the form of caves. These cavities can be the cause of earthquakes. The pulsating star reveals its own cavities not only in tremors, but noticeably in luminosity and temperature. And this is one more of the many phenomena that regularly occur in nature.

The Red Giant, Betelgeuse (millions of times the size of the Sun) is said to be a pulsating star. This strange feature may be most common among the Red Giants.

In revealed science, in the book of Enoch the Prophet, there are more than 1,000 classes of stars. How a star like the Red Giant or a regular sequence star like the Sun can become a White Dwarf or a Neutron Star may not be scientifically correct since there are more than 1,000 different kinds of stars. A White Dwarf could well be its original form, and not necessarily due to a change in the life of a Regular Sequence Star.

Planetary Nebulae

A Planetary Nebula is a massive spherical cloud illuminated by a star at its center. The Planetary Nebula, N G C 7293 is said to have a diameter of 2 light years and a central star one tenth the size of the Sun. It is believed that a Planetary Nebula represents a final stage of a Red Giant. This is merely an hypothesis, not an evidence. There are still many unknowns in our Universe. The central star of the nebula can contract into a Black Hole. It is believed that more than 60,000 planetary nebulae are located in the galaxy of the Milky Way.

Black Holes

No one can be sure about what causes a Black Hole but there are prevailing evidences in support of its existence. Most scientists seem to agree on the Theory of Black Holes. The collapse of a star under its gravitational force in which its Gravity overcomes its radiation is called a Black Hole. Black Holes are characterized by their heat and gravitational force, rather than by their masses. A Black Hole is extremely hot and its Gravity strong enough to trap its light within a very limited distance from its surface. Light travels at the speed of 186,000 miles per second. It is now trapped closely around the surface of the Black Hole. Scientists have failed to point out that Black Holes have exerted great gravitational force on neighboring stars. The velocities of these stars neutralize this force to the extent that they are not adversely affected.

The Theory of Black Holes is based on two main factors: the size of a star and its behavior after it has consumed its nuclear fuel. In his book, A Brief History of Time, Stephen Hawking points out that the idea of the Black Hole originated with John Mitchel in 1783. Mitchel's idea was that a massive and very dense star could develop such gravitational force that could severely restrict its radiation, so that from a distance its light would not be seen. Mitchel thought there were many such stars. This original idea has now been modified by other scientists.

A star more than twice the mass of the Sun, can contract into a Black Hole. A large star has more nuclear fuel than a small star but it consumes its fuel at a faster rate due to the greater degree of heat required to balance its gravitational attraction in order to radiate its light. The second factor is its behavior after its nuclear fuel is depleted. The Gravity of the star causes its matter to compress to an amazing degree. This causes the star to be much hotter than before. That is why a Black Hole is so hot and dense.

The attention of scientists is preoccupied with a Black Hole at the center of the Milky Way. They say this Black Hole has a mass of a million suns. It was featured in the science magazine, Discovery in the June issue of 1989. I quote: "The idea that our home galaxy has a Black Hole as massive as a million suns at its center – once considered a highly speculative notion keeps moving further into the main stream of astronomical thought…the core seems to be enveloped in a massive rotating shell of gas some 6-10 light years (about 50 trillion miles)."

The scientists felt that this shell of rotating gas was fueling the Black Hole but they could not figure out what was replenishing the shell. To them, this was a mystery. One scientist, Paul Ho of the Harvard Smithsonian Center for Astrophysics suggested there was a connection between the shell of rotating gas around the Black Hole, and a gigantic cloud of gas 20 light years away from it and cited evidence of connection —— a long thin stream.

I read this article several times and then began to think. Then in the fraction of a second the answer came to me. I reasoned, if a Black Hole was really what the scientists said that it was extraordinarily hot and dense with extreme gravitational force, and if outer space was 450 degrees below zero, then the heat from the Black Hole would interact with the cold and create the surrounding shell of gas. The distance of 6-10 light years would be just the appropriate distance for the clouds to form around the Black Hole.

In 1989, I had no idea of writing this book. When I received the answer to the surrounding clouds of the Black Hole, I realized then that I had a natural intuition for science. This gift of knowledge was always acting up and I was not really conscious of it. I remember as early as 1968 when I began pastoring my first church my intuition for science began manifesting. I was not married. The Church was not rich, so I had to reside with a family that was attending the church. They had a set of encyclopedias. I read science subjects and made notes. I have used some of these notes in this book. I was always curious about nature, I guess as everyone, but I did not study enough in science to be asking questions about the 23 degree tilted angle of the Earth, gravity, weightlessness of outer space, etc.

As you know, science is not the subject an evangelical minister or the subject a Theologian gives much of his attention. He scarcely gives it any of his attention. I think this is a great mistake of the Theologian. This was a great mistake on my part. God created the Universe; it reveals His glory. God created man with an infinite capacity for knowledge. Furthermore, our fore parents Adam and Eve chose knowledge to life. So man has a double endowment of knowledge. Knowledge is not without purpose. It is to be used to the glory of God. The study of nature is a noble purpose of the mind, but it should not be an end in itself. God is teaching about Himself through science, but man is not listening and learning. One can acknowledge this truth or ignore it. Ignoring it will not change the reality of it. If God had not given man knowledge, he would not be able to build scientific tools to discover and measure the Universe.

But there can be misinterpretations of God's revelations and in some cases absolute ignorance. In Theology it is chiefly misinterpretations. In science it is chiefly ignorance. The indispensable factor in all knowledge is God. And this factor remains interwoven with all of knowledge whether ignored, overlooked or acknowledged. God is the source of man's knowledge and it began when He breathed into his nostril the breath of life.

Some Tools Used by Astronomers

Before closing this chapter, let us look at some of the tools the astronomers use to study the stars and to measure the Universe. First is the Optical Telescope. Modern astronomy began January 7, 1610 with Galileo's 1 inch telescope. The telescope is designed to collect light from stars and focus it in a way as to give a picture of its source, by bending the light rays. There are two kinds of Optical Telescope: the Refracting and the Reflecting Telescopes. The Refracting Telescope uses lens to gather the light and bend the rays. When it bends the rays into a focal area, they cross over in the middle of the Telescope. These rays then enter the eye lens and produce a small picture which can be seen with the eye. Astronomers place a camera at the eye lens and photograph the stars.

The Reflecting Telescope uses a mirror and reflects the light rays into an eyepiece. The largest telescopes in the World are the Reflecting Telescopes because their mirror is easier to produce than the lens of the Refracting Telescope. The World's largest telescope is the Hale Telescope at Mount Palomar, California. Its diameter is 200 inches. This Telescope can intercept 40,000 times more light than Galileo's 1 inch telescope. Galileo's 1 inch telescope would only be able to show about 500,000 stars; the Hale Telescope shows billions of stars.

Other tools used by the astronomers are the Thermocouple, the Spectroscope, and the Radio Telescope. The Thermocouple is used with the telescope to measure the temperatures of stars. The Spectroscope is used with the telescope to study the elements of stars. The light rays are beamed into the Spectroscope. These rays pass through the Prism, a triangle made of glass. The different frequencies of light rays are separated. This is the spectrum of a star. By studying the spectrum, astronomers know the different elements in a star. The Radio Telescope is a most ingenious scientific tool. It receives the radio waves emitted by stars, amplifies, and records them. When the Radio telescope discovers a new source of radio waves (invisible rays), scientists use an Optical Telescope to find the source and to study it more carefully. A great advantage of the Radio Telescope is that it can be used in the day as well as in the night. This is why I have said God has revealed Himself in science by giving man the knowledge to build these scientific tools and that God wants us to have a true understanding of the Universe. Since it is obvious that God wants us to have a true understanding of the Universe, the Theologian should be able to formulate a theory of the formation of the Universe which could show how God formed it. Such a Theory has been formulated in Chapter 2.

CHAPTER 6
Motions of Bodies In Space

All bodies in space are in continuous motions. We are aware of two basic motions in the Solar System: Rotation and Orbit. We assume that these motions hold true of other bodies in space. These motions are not aimless but systematic and precise. This is possible by the weightlessness of outer space and by the laws of Gravity and motion. In this chapter the focus is on the origin of motions, the orbits of the stars, rotation of the galaxy of the Milky Way, and reconciling it with the Theory of the Expanding Universe.

The Origin of Motions

Experience has shown that Newton Mechanics (Theory of Gravity and Motion) has not explained the origin of the motions of the planets in the Solar System. Some planets that are larger and farther away from the Sun rotate at a faster speed than some planets that are nearer to the Sun. You can look back at the chart of the planets of the Solar System to see that this is so. Experience teaches that if you put an object into orbit of an attracting mass, whether it be the

Earth or the Sun, within a certain distance to the attracting mass and at a certain speed, that object will continue to orbit at that speed and at that distance. Thus we can clearly see that Newton Mechanics merely explains the systematic motions of a body, but not how those motions originated or were first established. So we are forced to accept that the motions of bodies in space and their speeds originated during their time of formation.

We have deduced from the scientific information in the Bible that the Universe was once in a state of dust and gas. From this state, let us visualize the formation of a galaxy of stars. God spoke and activate the laws of Gravity and Motion. Our picture of what happened next is perfectly clear. The Galactic Motions began. At the center of the Galaxy stars began to form. Clouds of dust and gas began forming into stars. Velocity of motions developed with the process of formation. The known velocities of the planets of the Solar System should give us an idea of what went on during the process of formation. We must understand that the formation of the Universe was not like the formation of an infant in its mother's womb. At God's command, matter began contracting at great velocities and under immense pressure of contraction.

This level of Galactic Activities was bound to produce variations: variations in speed of rotations and orbits, in sizes of stars, and in duration of formation. Some stars which began forming simultaneously with other stars would complete the process before others. There would be various groups of stars: from groups of two or more. After the formation of a group of stars, each member of that group would remain connected by gravitational attraction among that group. Their respective speeds of motions developed during the formation process would continue. The orbit of a body would be limited to the circumference of the volume of dust and gas from which it was formed.

The Solar System, like any constellation of stars, was formed from clouds of dust and gas. Its formation began simultaneously. After its formation, it remains connected, the Sun being the center of Gravity because it contains more than 98 percent of the matter. A constellation of stars shall remain gravitationally connected so long as the Universe shall last. Some scientists suggest that the rotation of the Galaxy will cause constellations of stars to spread out and finally cease to exist as constellations. "Rotation of this galaxy will, therefore, in the course of some tens of millions of years, cause every stellar association to spread out in the direction of the rotation and assume an oblong shape. The fact that associations expand means that their life as groups is relatively short; in a few million years they will have become completely unrecognizable as such." <u>Atlas of the Universe op. cit p.99</u>

Here we have an instance when modern science contradicts the scientific information of the Bible. The Bible is very clear and emphatic about the continuity of the constellations. Here the words of the Creator: "Canst thou bind the sweet influences of Pleiades, or loose the bands of Orion? Canst thou bring forth Mazzaroth in his seasons? Or canst thou guide Arcturus with his sons? (Job 38:31, 32). When God speaks in such a clear and emphatic manner on a subject of science, then His words must be accepted above the words and scientific predictions of the scientists. The science of Cosmology and Astronomy are sciences in which scientists are bound to make errors.

The Orbits of Stars and the Sun

Astronomers have reported that about 20 percent of the visible stars orbit in groups of two, two stars orbiting each other. In some instances the orbits vary from 3 to 45 years. In most cases one of the stars is much larger than the other. Astronomers have even reported instances in which a star was seen orbiting an invisible companion believed to be a Black Hole.

While astronomers have identified the proper orbits of 40 percent of the visible stars, and also the proper orbits of stars in various constellations, they have failed to identify the proper orbit of the Sun – the closest center of Gravity the Sun is orbiting. In many science books there is no mention of a proper orbit of the Sun. These books simply suggest that the Sun is orbiting around the center of the galaxy of the Milky Way. In such cases the Sun takes some 220 million years to make one revolution. The distance from the center of the Galaxy determines the time period of the orbit.

One cannot ignore the fact that the relatively close proximity of the Sun to us presents some difficulty in identifying its proper orbit. If the Sun were 5 light years away from us it would be much easier to determine its proper orbit as some of the stars which are seen orbiting in groups of two are about that distance away. But a scientific principle is only true if it applies without restrictions: it must equally apply to large as well as small, near as well as far. If there is a principle that regulates the orbit of a body, one does not necessarily have to observe a particular orbit in order to identify it.

Newton's first law of Gravity that everybody in the Universe attracts every other body implies that a body will orbit the closest center of Gravity. This first law is graphically illustrated by the planets of the Solar System and by the observation of two stars orbiting each other. The structure of the atom in which electrons orbit around its center also illustrates Newton's first law of Gravity. Accordingly, one should expect the Sun to obey this law. To say that the Sun orbits around the center of the Milky Way is the simple and convenient thing to do and is admission of failure in science to identify the closest center of Gravity the Sun is orbiting.

In his book, The Meaning of Relativity, Einstein was careful to point out that E. Mach's attempt to eliminate space as an active cause in the system of mechanics by suggesting that a material particle does not move in uniform motion relative to space but to the center of all other masses in the Universe. Einstein was also careful to point out that E. Mach had failed in his attempt. It is clear that the orbit of the Sun as explained by scientists is based on Mach's condemned theory. Applying Mach's theory to the motions in the Galaxy of the Milky Way means that the Sun would not orbit the closest center of Gravity, but instead would be orbiting around the center of the Milky Way.

This is only part of the fact of the Sun's orbit and not the greater part of that fact. Indirectly, the Sun orbits around the center of the galaxy but not directly. My opinion is that the Sun obeys Newton's first law of Gravity. It is orbiting the closest center of Gravity. The nearest stars to the Sun are within the radius of 4 to 11 light years. There are only 11 of these stars. Nearest of them is Proxima Centauri, at a distance of 4.2 light years. Proxima Centauri is a member of the three star constellation of Alpha Centauri. It is 0.1 light year nearer to the Sun than either of the other two stars in the constellation. The Sun makes its yearly orbit close to the 11 nearest stars, of which it seems to be a constellation member. Alpha Centauri is included in the 11 nearest stars to the Sun

The constellation of Alpha Centauri appears to us as one star to the naked eye, and thus the third brightest star in the sky. These stars are said to orbit each other every 80 years. Proxima Centauri in its orbit around the other two stars comes to us closer than any of the other two stars. Alpha Centauri is said to be approaching the Solar System at 15.5 miles per second. It is predicted that in 28,000 years it will only be 3.1 light years away.

This scientific information seems to suggest that the Sun is the most remote member of the Alpha Centauri constellation. Instead of a constellation of three stars, apparently it is a constellation of 4 stars. An observation of the motions of both the Sun and Alpha Centauri from a distance of one light year away would prove that the Sun is a member of that constellation. Thus the proper orbit of the Sun would have been identified. The "Ecliptic" the apparent yearly path of the Sun among the twelve constellations of the "Zodiac" makes better scientific sense than the idea that the Sun is orbiting around the center of the Milky Way. Until 1992 I thought that the Sun was orbiting around Alpha Centauri. But in 1993 my position was changed and I then thought the Sun was orbiting within the limits of the 11 surrounding stars. After considering the eleven stars surrounding the Sun from a distance of 4 – 11 light years, I realized that the Sun has more than enough space in which to complete its yearly orbit. Having understood that its orbit is very much restricted, compared with other stars, I had no choice except to abandon my deduction of 1992. It will be shown that the Sun's orbit is unique and that it is obeying a direct Divine command.

Rotation of the Galaxy

Having said that the Sun's orbit is unique, obeying a direct Divine order, let us now turn our attention to the rotation of the galaxy. The Gravity of a mass converges in or pulls towards its center. This is why the matter of a mass is held together. This principle also applies to the mass of the Milky Way. The rotation and orbit of every star in a galaxy are inclined towards its center. Stars at the center will naturally be older than other stars and may be more massive. These stars are more likely to collapse into Black Holes. Scientists have reported the existence of a Black Hole at the center of the Milky Way whose mass is more than a million times that of the Sun. It is thought that there are much larger Black Holes at the center of other galaxies.

In the St. Petersburg Times (Florida, USA), July 10, 1992 was the Head-line: ASTRONOMERS FIND BLACK HOLE, BELIEVED TO BE LARGEST EVER.

A swarm of ordinary stars buzzing about the dark and unseen center of a distant galaxy has helped astronomers locate what is a massive black hole ever discovered. John Kormendy of the University of Hawaii Institute of Astronomy said Thursday that the Black Hole fills a volume about equal to the Solar System, which includes the Earth and Sun. It has a mass equal to about a billion suns. The Black Hole is in Galaxy NGC3115, a stellar grouping that previously was thought to be only a cluster of ordinary stars some 30 million light years from Earth.

A scientific interpretation of the article is that Black Holes serve as Gravitational Centers in their respective galaxies. Therefore Black Holes are more critical to the galaxies than once thought. Let us now picture the gravitational attraction of the Milky Way which is the collective gravitational attraction of the individual stars. This is what holds the galaxy together. When we apply E. Mach's Theory to the galaxy, it means that the motions of the stars are inclined towards its center, but the Theory does not recognize local centers of Gravity within the Galaxy. That is why I have said the Theory represents a part of a fact and not the greater part of that fact.

Since astronomers have reported star systems in which two stars orbit each other and systems in which three or more stars orbit each other, local gravitational centers within the galaxy have been established. This means that although the principle that the motions of all stars are inclined towards the center of the Galaxy, these stars are not directly orbiting around its center, only indirectly. From the mere fact that all these stars are located at various distances all around the center of the Galaxy, they are indirectly orbiting its center.

We can visualize the gravitational center of the Galaxy as a core of stars quite possible the most massive stars orbiting each other. A number of other stars may be directly orbiting around each other. A number of other stars may be directly orbiting around this nucleus. If the information about the Black Hole at the center of the Galaxy

Footnote: The reader is hereby referred to the Illustrated Encyclopedia of Astronomy and Space, published by Thomas Y. Cromwell NY, 1979 Ed. for further details of the constellation of Alpha Centauri.

is correct as to its gravitational attraction and mass, it forms a part of this nucleus. One may deduce that this Black Hole is only rotating. The stars of this nucleus would orbit around it. The construct, that the orbit of a star would be within the circumference of the volume of dust and gas from which it was formed, makes it impossible for very many stars to directly orbit around the nucleus of the Galaxy.

Scientists suggest that in hundreds of millions of years the stars in the galaxy will eventually orbit around its center depending on their respective distances from it. It is very convenient to say this as it is the popular thing. The idea of the rotation of the Galaxy is simple. Stars have two basic motions. They rotate and move in a circular line within various groups. Since the stars in the Galaxy sustain these basic motions, it must be understood that the Galaxy is rotating. This simple and logical way of explaining the rotation of the Galaxy is not what is meant by the following authors:

We live in a galaxy that is about a hundred thousand light years across and is slowly rotating; the stars in its spiral arms orbit around its center about once every several hundred million years. Our Sun is just an ordinary, average-sized yellow star near the inner edge of one of its spiral arms. A Brief History of Time op. cit. p.37

It has also been found that the Galaxy is rotating around its center; the Sun takes 225,000,000 years to complete one revolution-period which is officially called the cosmic year .The Amateur Astronomers' Glossary by Patrick More p.50. Rotation of this Galaxy will, therefore, in the course of some tens of millions of years cause every stellar association to spread out in the direction of rotation and assume an oblong shape. Atlas of the Universe op. cit p.99.

The rotation of the Galaxy as explained by these authors cannot, to say the least, be supported by any scientific evidence. It is based on E Mach's Theory which had as its premise the elimination of space as an active agent in the system of Newton Mechanics. A scientific theory should help us understand the Universe. This Theory given us by the scientists has not done so. Therefore, it must be rejected and removed from all Science Books in the hope of providing a better understanding of the Universe.

Reconciling the Theory of the Rotation of the Galaxy with the Theory of the Expanding Universe

Scientists explain the Universe is expanding; galaxies are racing from each other. At the same time they explain galaxies are rotating. Can the two be reconciled? Let us give this our attention. In our review and response to scientific theories in chapter 3, the Expanding Universe was discussed. The substance of the Theory is that galaxies at the outer reaches of the Universe are receding from each other at incredible speeds; the farther they are away, the faster they are moving. This observation forms the foundation of the theory. Part of our response to the Theory, there and then, was that if one were to view the center of the Universe (looking from the outer edges means looking at the center) from the outer edges, one would have seen the galaxies at the center spreading farther and farther apart with similar velocities. So how could one reconcile the Theory of the Expanding Universe with the Rotation of the Galaxy?

The expansion of the Universe would have very serious consequences for its stability. First of all, it would have caused the constellation of stars to spread out and not the so called rotation of the Galaxy as some scientists suggest. Orbits would be affected by any meaningful expansion of the Universe. Space represented dark invisible matter. This created condition of weightlessness. The idea is that a galaxy has more invisible matter than visible matter. This supports the weightless condition of a galaxy. If a galaxy keeps spreading apart from the rest, very soon it would collapse because it would be losing the over-all support of the invisible matter, which in the first place provided the weightlessness of outer space It is like the foundation of a house which keeps rending apart. Very soon parts of it will begin collapsing. Before long the whole house will collapse. It does not really make sense in talking about the rotation of a galaxy when the conditions of rotation are absent. Does it?

A reconciliation between the two theories is just not possible. An assumption or theory can describe or even predict a reality but an assumption and a reality are not always the same thing. At times an assumption can be remote from reality as one can imagine. And we here have a classic example before us. Since one cannot reconcile the Theory of the Expanding Universe with the Theory of the Rotation of the Galaxies, the reality is that neither the Universe is expanding nor the galaxies are rotating the way scientists explain. One obstacle to a true scientific description of the Universe is the self-contradictions of the scientists. A true scientific description means the elimination of all such contradictions.

Everything in the Universe is moving at incredible speed because of its weightless condition. But its weightless condition is not random. It has to be proportional to its masses. A fraction of imbalance would create a Cosmic Storm in which all masses of matter would be tossed to and fro like the raging waves of the sea. Orbits and rotations could not be predicted. Earth would not be an exception because we would not be able to predict time of rotation and its orbit around the Sun. The difference between an earthly storm and a Cosmic storm is that there would be no end to the Cosmic Storm. By this token, we know that there had to be a CREATOR.

CHAPTER 7
The Economy of Cosmic Radiation and Recycling

Like everything else, the Economy of Cosmic Radiation and Recycling is in-wrought within the fabric of the Universe, while being intertwined the one with the other in a manner as to form a "Cosmic Compound". A compound can be used for a single purpose. On the other hand, it can be broken down into its elements and used for different purposes. Here we have chosen to break down this compound and treat its elements accordingly. Cosmic Radiation is the radiation of all radiating bodies in the Universe. Cosmic Recycling is the recycling of matter, including radiation, in various forms in the Universe. We on Earth are affected in every area of life. But it must be emphasized that Cosmic Radiation and Recycling are efficiently managed. And it is upon this efficiency of management that our life depends and life in its simplest form. It is truly amazing that in an infinite Universe there is no place or time for cosmic waste. The countless stars including our Sun convert countless tons of matter into radiant energy every moment of time, which shines through the darkness. Still none of this energy is wasted.

Radiation of the Sun

The Sun converts about 4 million tons of matter into energy every moment of time. Through the process of Nuclear Fusion, this energy is radiated into space. The Sun's radiation cycle seems to have prevented any noticeable loss to its mass. Of this radiation, only a small percentage reaches the Earth. The greater portion travels through the Solar System and to outer space. Most of the radiation that reaches the Earth is absorbed by the Atmosphere, the oceans, and the Earth's surface. The Earth's surface absorbs much of this radiation. Having laid the background of the Sun's radiation, we can now look more closely at its economy, its management.

The Sun's radiation is critical to the maintenance of life conditions and all life forms here on Earth. But there is no element of nature independent of the other. The Sun needs the Atmosphere in this critical role. The absorption of the Sun's energy by the Atmosphere converts it into heat. The Atmosphere spreads this heat around the Earth. This helps to balance the Earth's temperature. Due to the 23 degree tilted angle of the Earth, the Equatorial Regions receive more of the Sun's radiation than the regions of the North and South Poles. This causes the Atmosphere in the Equatorial Regions to be much warmer. But the continuous movements of air masses spread the warmer air causing a more balanced temperature.

The interactions between the Sun's radiation and the oceans are also important to the maintenance of life. When the oceans absorb the Sun's radiation, it is converted into heat. This heat causes billions of tons of water to evaporate daily into the Atmosphere which is circulated over the Earth. Sometime after evaporation, the water returns to Earth in the form of rain upon the oceans and dry land, but not before a chemical process occurs. Without the Sun's radiation, the Earth's water cycle would not have been possible. The Bible speaks about this phenomenon—"It is he that builds his stories in the heavens and that hath founded his troops in the earth; he that calls for the waters of the sea, and pours them out upon the face of the earth: The Lord is His name", Amos 9:6. We do not hear God's voice but we see when He pours the waters of the oceans upon the face of the dry and parched land.

Modern science teaches that the Sun's radiation interacts with the Atmosphere and oceans causing winds, rains, lightning, and thunder. Long before the dawn of modern science the Bible clearly describes the Sun's interactions with these other elements of nature: "Out of the South cometh the whirlwind: and cold out of the North. Do you know the balancing of the clouds, the wondrous works of Him which is perfect in knowledge? By what way is the light parted which scatters the east wind upon the earth. Who hath divided the water course for the overflowing of waters, or a way for the lightning of thunder; to cause it to rain on the earth..." (Job 37:9, 16; 38: 24-26).

The Sun's radiation is not only critical to life conditions but also to life itself. All animal life and all plant life directly depend on it. The ant is one of the smallest insects and, from a human point of view, insignificant, but like every other living creature utilizes the Sun's radiation. The tender plant by its leaves, through the process of photosynthesis, converts the Sun's energy into food. Thus we have seen how the small percentage of the Sun's radiation, reaching the Earth, has been used.

The greater percentage of the Sun's radiation goes into outer space. We should not assume that it is a woeful waste. Rather we should assume it goes towards the sustenance of the Universe in other areas. The Universe is a marvelous system. There are many critical parts to a system. One of the critical parts in the system of the Universe is its Magnetic-field. In discussing Newton's theory of Gravity and Motion in chapters 3 and 4, it has been pointed out that Newton's theory of Gravity and Motion was not a complete theory and a number of observations made in the hope of a better understanding of those laws.

One of the Theo-Cosmo observations of Newton Mechanics is that Gravity works through the Magnetic-field of the Universe. It now becomes necessary to here explain more about it. Most of cosmic radiation is channeled into the Magnetic field of the Universe. For the most part, it is undetected. Yet there are evidences of it. The Microwave Radiation is one such evidence. Microwave Radiation fills the Universe. It consists of very short radio waves at cold temperature of about 454 degrees below zero. Some scientists suggest that Microwave Radiation is the glow from the early Universe.

I cannot share such scientific assumption. As a matter of fact the scientific evidence points to the contrary. If we accept the Universal Law of the Conservation of Energy that energy cannot be created nor destroyed, only transferred or transformed, the radiation from all bodies in space must remain in the Universe in one form or another. For the most part such radiation remains undetected and invisible. There is more significance to Cosmic Radiation than we will ever know. We should endeavor to know as much as is possible. Man has an infinite capacity for knowledge.

Radiation of Stars and the Magnetic Field of the Universe

It is not impossible for human beings to number the stars. We can only have a faint idea or very rough estimate to say the least. It is also impossible to estimate the energy generated by all the stars. We know that most of them are Regular Sequence stars like our Sun. The Red Giants, the largest stars in the Universe, each generates an enormous amount of energy many times greater than that generated by the Sun. What has happened to all of this energy? Based on the Universal Law of the Conservation of Energy, the Economy of Cosmic Radiation and Recycling implies that most of the radiation is collected into the Magnetic-field of the Universe. By the motions of stars in groups, as galaxies, and as distinct units, energy is equally distributed throughout the Universe, maintaining its Magnetic-field. The Magnetic-field in turn affects the motions of stars through the Laws of Gravity and Motion and also connects all galaxies so that each galaxy is inter-related with the other. Galactic Gravity is only a greater degree of the function. And in this function, it is as effective as it is in the Solar System and in Earth's own Gravity.

A principle is only a principle if it is unrestricted in its application; it must apply without limitation. It was with this understanding that the suggestion was earlier made that our Sun must orbit the closest center of gravity. However, the Sun's orbit is unique: it is obeying a direct Divine command. It Chapter 8 the orbit of the Sun is fully dealt with and clearly shows the Suns exception. In the Theo-Cosmos Theory of the formation of the Universe, it was pointed out that the Nuclear Fusion of a star began during its formative state. A star lives by burning. The early Nuclear Fusion in its center of clouds of dust and gas would speed its development. As all the laws were intertwined with the formative state of the Universe, all physical phenomena, including the Magnetic-field of the Universe, originated and developed. As Hydrogen Clouds of dust and gas contracted, compressed, and rotated at incredible speed, cosmic radiation filled the Universe. This radiation formed into the Magnetic field. With a fully developed Magnetic-field and the development of stars, the Universe is now experiencing its fine-tuning. All the motions of the starts are now fully established through the Magnetic field. And by distance and velocity, the attraction of Gravity is balanced. The result is that motions can be calculated and predicted with absolute precision.

Cosmic Clouds

During the process of formation of a star, some of its matter escaped and remained cosmic clouds. The amount of matter in the form of clouds, is phenomenal. However, some of this matter is the result of Cosmic Radiation. The most of the matter in the Universe is cold invisible matter. This was earlier explained in chapters 1 and 4. The Universe is colder than it is hot. What would have happened if it was hotter than it was cold? I am sure I would not have been around proposing such question. It is obvious that the **cold invisible matter** absorbs a great amount of radiation. The interaction between radiation and cold invisible matter is similar to that between the Sun's radiation and the oceans. The oceans absorb a great amount of the Sun's radiation but billions of tons of water are evaporated into the Atmosphere daily. In like manner cosmic radiation causes Cold Invisible Matter to form into clouds. Through the work of Gravity and Motion some of these clouds are positioned to become luminous clouds. Other clouds are so gigantic and so positioned that while absorbing radiation, still remain dark and prevent such radiation from spreading. These clouds can be the source of new stars.

Because energy cannot be destroyed, we can assume that the radiation of stars takes on other forms. It can interact with cold invisible matter to form other gases. It can also illuminate other clouds which will reflect light. Outer space is the ground for many ACTIONS, REACTIONS, AND INTERACTIONS undiscovered and unknown to man. After all, Relativity may not be all that complex and intricate. It is only that complex as we would want to make it because all atoms, molecules, and elements are interdependent: they altogether make the Universe what it is The ONENESS of God is clearly reflected in all things.

Cosmic Recycling

Cosmic Recycling is a daily process. The recycling of matter takes place throughout the Universe. By this law the Universe could last forever. Sin is the only reason that the present Universe will not last forever. The eternal and infinite nature of God dictates that the Universe exists eternally. There must now be a physical representation of His infinite, omnipresent nature. God's original purpose in Creation will not be thwarted by sin. So He will create another Universe, one that will be free from sin and will last eternally. The creation of a new Universe is the fundamental difference between natural and the spiritual reality. As long as there is a natural Universe, the law of Cosmic Recycling will apply.

Since Cosmic Recycling is a law of nature, let us briefly reflect on its occurrence here on Earth. It is in all aspects of life conditions and life forms. Matter exists in three basic forms: solid, liquid, and gas. A solid form of matter can be changed into the form of gas. The most simple and graphic illustration of this is the change of ice into water. The water can be easily changed into gas by heating it on your stove. Even iron can be changed into liquid form and if further heated, can be changed into gas. Gas can be changed into liquid and be further changed into a solid form. In the water cycle, the ocean water evaporates into the Atmosphere and then falls as rain. The change of one form of matter to another and the reverse process completes that cycle.

When a warm air mass takes the place of a cold air mass, circulates and becomes a cold air mass, that air cycle is complete. When the plants convert the carbon dioxide into oxygen we inhale it and when we exhale it, it again becomes carbon dioxide. At this point the cycle is complete. When plants and animals reproduce themselves, those cycles are complete. Modern science and technology have brought about our present industrial revolution in which recycling is a daily occurrence in our domestic and national life. There are numerous products of a recycled nature particularly in paper, plastic, and aluminum products. This recycling revolution is a prototype of that which occurs in nature.

Radiation Cycle: Nuclear Fuel

For some time, before I began writing this book, I could not escape the idea that there was something more about star radiation; some secret that the scientists have not yet discovered. Then the idea of a radiation cycle had impressed itself upon my mind. The idea is that the Sun produces its Nuclear Fuel through its radiation cycle. This idea stands in direct contrast to the scientific theory that within 5,000,000,000 years the Sun will have exhausted its Nuclear Fuel. The Sun was created to manufacture its nuclear fuel which it has been doing since Creation. The Sun's inner core sustains a temperature of 20,000,000 degrees Centigrade. At this temperature Hydrogen is produced, then rises to the upper layers of the Sun where they fuse to produce radiant energy. The Sun's inner core is the key to its radiation cycle The Hydrogen atoms fuse to produce Helium. In the process, radiant energy is released.

We must now inquire as to how the Sun's radiation cycle is sustained. There are three known layers of gases of the Sun: the Photosphere, the shinning disk; the Chromosphere lying above; and the Corona lying above the Chromosphere. These layers of gas range in temperature from 6,000 degrees Centigrade to 1,000,000 degrees Centigrade with depths from 10,000 to 1,000,000 miles. The Photosphere is a mirror of what of what is happening in the Sun's radiation cycle. In the Photosphere Hydrogen Atoms fuse to produce Helium resulting in radiant energy. All three layers of gases are held together by the Sun's Gravity from which they can never escape; all that can escape is its radiant energy. With the Sun's Gravity and with its inner temperature of 20,000,000 degrees Centigrade, these layers of gas interact to sustain the Sun's radiation cycle.

We are no strangers to Gravity; we know what it can do and we know that the Sun's Gravity is many times greater than that of the

Earth. The Earth rotates at 1,000 miles per hour, yet its Gravity keeps its Atmosphere around it. Similarly, the Sun's Gravity holds its layers of gases together, sustaining its Radiation Cycle. Our friends, the scientists, tell us that most of the stars like the Sun will have exhausted their nuclear fuel within 5,000,000 years; yet they have failed to make any prediction of the Earth's water cycle, which logically would be more precarious than the Sun's Radiation Cycle.

Star radiation is not a one way street any more than the Earth's water cycle. There is always an invisible aspect to nature. And this is what makes the study of the Universe the most profound challenge to the human mind; and on the other hand, the greatest revelation of a Creator. Billions of tons of water are transported into the Atmosphere daily through the process of evaporation. With the human eye we do not see this aspect of the water cycle; yet we see the rain as the water returns to the oceans. The Sun's Nuclear Fusion is sustained by its radiation cycle and is guaranteed by its inner core and gravity. The phenomenon of the Radiation Cycle is one of the marvels of nature. Now, we do not have to be concerned about the depletion of the Nuclear Fuel of the stars. If the scientists are not concerned about the Earth's water cycle, they should not be concerned about the stability of the stars.

The Gravity of the star is a key factor in its Radiation Cycle, but there are other factors that are equally important in the process. The study of nature is, indeed, a noble exercise of the mind. The primary purpose of the Sun or the stars is to radiate light in a Universe of darkness. Scientists say when a star's inner structure contracts to certain degree or becomes super-dense, it may contract into a White Dwarf, or a Black Hole, or it may even explode. These are the exceptions. Because the Universe is measureless, it can accommodate such phenomena without any damage to its stability.

God's original purpose in Creation was that the Universe should last forever. And in creating matter, God also created those laws that would ensure His purpose. Prominent among these laws is the Law of Cosmic Recycling. Cosmic Recycling includes radiation recycling. If no law of radiation recycling existed, the Universe would not have lasted 1000 years. Scientists inform us that the Sun has enough Nuclear Fuel that will last another 5 billion years. Let us add another 5 billion years to this and say 10 billion years. What next? This means that by this time all the stars would have exhausted their nuclear fuel, many of which would have exploded.

Indeed very little would have been left of an infinite Universe. And God would have begun a new work of Creation. This He would have to do every 20 billion years. Certainly He has His work cut out for Him. Does not He? For us human beings 20 billion years is like the whole of eternity. But to the Eternal Being, 20 billion years is like a moment in eternity. Simply, I do not think God has any interest in creating a Universe every 20 billion years or so. God's original and eternal purpose in Creation SHALL STAND. His Universe shall last forever. That is why the law of Cosmic Recycling is so fundamental to our understanding of the Universe, particularly the radiation cycle of stars. Indeed, only a small percentage of the 4 million tons of matter converted into energy every second by the average star is ejected into outer space in the form of light. Most of it is retained in the Corona and the Chromosphere which are critical to the Radiation Cycle.

Recycling in outer space may take several forms. Over a period, radiation itself can take several forms. **Cold invisible matter**, after interacting with Cosmic Radiation, can recycle into clouds. When stars explode, they become clouds of dust and gas; and when such clouds of dust and gas become stars, that particular Cosmic Recycling is complete. Star explosion is another means of Cosmic Recycling An ordinary star explosion is called a nova; others are called Supernovae. These are less frequent. However, more than six occurred during the last thousand years. As recent as February 1987 a star exploded in our galaxy, the Milky Way. Some stars are many times larger than the Sun. The explosion of a star near the Solar System could destroy all life on Earth.

Scientists say a star explodes when its Nuclear Fuel is exhausted depending on its size. The exhaustion of Nuclear Fuel in stars the size of the Sun can result into such stars as White Dwarfs and even Neutron stars. The exhaustion of Nuclear Fuel in stars larger than the Sun, depending how large, is called a nova or supernova. Most of this matter is recycled into other forms, for example, into Hydrogen Clouds. These clouds can be come luminous soon after.

What can cause a star to explode, scientists explain, after the exhaustion of its nuclear fuel is that its core contracts to such a degree that its outer parts begin to collapse. The collapsing parts interacted with the contracting core causing extreme temperatures. This caused the star to be blown apart. While this phenomenon is rightly called supernova, it is not an indication of the end of the Universe. If a Neutron star resulted from this phenomenon, most of the matter of this once large star returned to space and intermixed with other clouds of dust and gases. If a new star is later formed from this, recycling occurs.

Stars have exploded but for reasons other than what is given by scientists. They have not made a research on the more than 1000 classes of stars. As a matter of fact, scientists are not even aware that there are so many kinds of stars. They have only mentioned five kinds of stars. The particular kind of star had everything to do with its explosion, which may have had nothing to do with an exhaustion of Its nuclear fuel.

Astronomers have explained that before some stars exploded they acquired extra brightness and then exploded. In the Andromeda Nebula, which has been interpreted to be an island Universe, astronomers discovered in 1885 a new star. Its brightness was calculated to be one-tenth the brightness of the entire nebula. In 1917 they found two other new stars of exceptional brightness. These three stars were considered to be the initial stage of a star explosion. Stars that are formed by the Law of Cosmic Recycling may have taken many years to form. This time period stands in direct contrast to the time period in the creation of the original stars. The Divine command of Creation would speed the formation of the original stars. It took God just one day to create the first man but it took the biological laws nine months, and another 20 years before that child becomes a man. This must also be true of Cosmic Recycling as it relates to star formation.

By the nature of Cosmic Recycling herein addressed we are impelled to draw a fine line between it and the Theory of the Expanding Universe. Cosmic Recycling in no way implies any form of expansion. There may be more plants and animals today than when God first created the Earth. But that did not mean the Earth was expanding. Many stars may have formed by Cosmic Recycling. At the same time, many stars may have exploded. Notwithstanding, this cannot be regarded as an expansion of the Universe. And as it has already been pointed out in chapter 6, spatial expansion has serious consequences for the stability of the Universe.

One amazing thing about the Economy of Cosmic Radiation and Recycling is that they both reflect an equilibrium that is truly amazing. Radiation is balanced by absorption; the sources are adequate for the cycles. In our Universe nothing goes to waste, absolutely nothing. This is further evidence that the Universe could not have accidentally evolved.

The greatest number of stars are Regular Sequence Stars, like the Sun. One must assume that they have similar layers of gases. As we have already pointed out, these gases sustain the Radiation Cycle within the stars. Considering that the Universe was created to sustain itself, this cycle is indispensable. Scientists overlook this critical factor and suggest that over a period of some 5,000,000,000 years they will deplete their Nuclear Fuel. However, these layers of gases are not only indispensable to Star Radiation Cycle but also to their motions in conjunction with the weightlessness of outer space. These gases being held by Gravity give momentum to the stars. .

CHAPTER 8
The Orbit of The Sun

After writing this book 1993, I had written a number of publishers in an effort to have it published. It is now May 1997 and I have not yet succeeded. However, one good thing had happened between then and 1997——it is that I have been able to identify the location of the Sun's orbit. In chapter 6 the ridiculous scientific theory that the Sun is orbiting around the center of the Galaxy has been struck with a death blow, thus abolishing it from any and all scientific considerations.

Until 1992 I thought that the Sun was orbiting around Alpha Centauri. But in 1993 my position was changed and I then thought the Sun was orbiting within the limits of the 11 surrounding stars. After considering the eleven stars surrounding the Sun from a distance of 4 – 11 light years, I realized that the Sun has more than enough space in which to complete its yearly orbit. Having understood that its orbit is very much restricted, compared with other stars, I had no choice except to abandon my deduction of 1992. It will be shown that the Sun's orbit is unique and that it is obeying a direct Divine command. By observing the motions of the Sun, we should have already concluded that its orbit is very restricted. However, Patrick Moor in his book "The Amateur Astronomers Glossary" did not help us to better understand the Sun's orbit. Defining the Sun's orbit, he states, the Ecliptic is the apparent yearly path of the Sun among the stars passing through the twelve constellations of the "Zodiac". But in His book, Stars and Planets" Joachim Ekrutt states: "The Sun seems to move and appears in front of more northerly constellations and other times more southerly ones. This apparent circular path the Sun takes across the sky in the course of a year is called the Ecliptic". Indeed, Joachim Ekrutt's description implies a more restrictive orbit.

The Ecliptic does not mean that the Sun actually passes by the twelve constellations of the Zodiac (animal sign of the constellations). In reality that is impossible. These constellations are many light years away. The laws of science do not allow for any material object to travel at and above the speed of light — only light itself. Were the Sun to travel at the speed of light, it would take more than four light years to reach Alpha Centauri, the nearest constellation.

As the Sun moves in its yearly orbit and as the Earth orbits around the Sun, the Sun appears to have passed by the twelve constellations of the zodiac because of its distance to us, size, and brightness. The Sun's brightness prevents us during the day from seeing the lights of the immediate neighboring stars and many others as well. You can easily see the effect of the Sun's brightness. Its brightness extends billions of miles into space.

During the Earth's yearly orbit of the Sun, the heavens are seen from different positions. The Earth also is seen in varying positions, in its orbit, during the four seasons of the year. The Sun also is in varying positions in its own yearly orbit. In all of this, the Earth's motions become a critical factor because the Earth not only rotates and orbits around the Sun, but it also follows the Sun as it sweeps through space in its orbit. We do not see the Earth moving. However, the Earth is moving across the sky in a circular manner, following the Sun.

As the Sun moves in varying positions in its yearly orbit, it eclipses the light of the constellations; and as the Earth orbits the Sun in varying positions, it appears as though the Sun passes through the Twelve Constellations of the Zodiac. Having eliminated from all scientific considerations the theory that the Sun is orbiting around the center of the Galaxy of the Milky Way and having explained the Ecliptic, the question is what then, if any, is the location of the yearly orbit of the Sun?

In his compendium of the laws of the Sun, the Prophet Enoch brings to light revealed science which must now impact modern science with astonishing accuracy. The renowned Prophet claims his compendium of the laws of the Sun was given him by the Arch Angel, Uriel and he wrote as he was commanded. Included in his compendium are a computation of the days of the year, the monthly and quarterly circuits of the Sun's yearly orbit, the coordinates of the Sun's motions, and the point and time of beginning of the Sun's orbit. And to conclude his compendium, the Prophet states unequivocally, "The Sun turns back twice in its yearly orbit". We must now apply this compendium to the reality of nature.

Application:

The Sun completes a total of 364 circuits in its yearly orbit. The Sun does not orbit in a 360 degree circle. Instead it orbits in a series of 364 daily circuits. It rises in the East and sets in the West. When it returns to the East a 24 hour circuit is completed.

Monthly and Quarterly Circuits

The monthly and quarterly circuits are constant. Each month has 30 circuits (30 days). There are 12 months, 360 days. There are 4 quarters. Each quarter has 90 days, 360 days. At the end of each quarter an extra day is added for the separation of the seasons, making 364 days. This means that 4 of the months have 31 days and 8 of the months have 30 days each, a total of 364 days in the year. Modern science computes 365 ¼ days in the regular year and 366 days in the Leap Year. Every 4 years an extra day is added. Revealed science can have no error. So the 364 day year must be the correct computation.

The Co-ordinates of the Sun's motion. There are 12 Co-ordinates: six of them in the East and six in the West. A set of Co-ordinates mark the volume of space in which the Sun completes its daily and monthly circuits.

The Beginning of the Sun's orbit—point and time. The point of the Sun's orbit has to do with its position in its orbit and in relation to other bodies in space. If we know the present co-ordinates, we can determine what time remains for the completion of its yearly orbit. But we must first know the point of beginning. One revolutionary aspect of Enoch's compendium is that the Sun begins its yearly orbit in the Forth Co-ordinates instead of the First Co-Ordinates. You will see that the Fourth Co-ordinate corresponds to the month of April. April is the Cosmological beginning of the year. Israel was commanded to observe Abib as the beginning of their year. [Exodus 12 and 13:4] Was this a mere coincidence? God wanted Israel to begin their New Year according to His Divine Calendar. Is it a coincidence that the Christian religion begins in April with the Death and Resurrection of Christ. And is it co-incidence that nature springs forth in April from the bondage of Winter? April is a special month in the Divine calendar. And so it is special to Christians. The Hebrew word, Abib means new, beginning.

The Sun's Orbit: The First Quarter

The Sun begins its yearly orbit by rising in the East in the Fourth Co-ordinate and goes South West to the Fourth Co-ordinate in the West. Then travels from the West north-east to the Fourth Co-ordinate in the East. This completes its first circuit [24 hours]. This is the first day of the first month, April first. The Sun is going North from this point and from this day. But in order to go North, the Sun must make a north adjustment in its angle in the West Co-ordinate. Without this adjustment, the Sun would remain static in the circuit of the first day and thus would not be able to go North-ward to the Fifth and Sixth Co-ordinates.

This Northern angle adjustment in the West will take the Sun North to the Fifth and Sixth Co-ordinates. While this Northern angle adjustment is fundamental in the sense it will take the Sun North, it does not fundamentally change the width and length of the daily circuit. Notwithstanding, there will be slight variations. This will cause some days to be longer [more day light] and some nights to be shorter. By the way, the lengthening of the day by the variation does not affect the 24 hour rotation of the Earth. It only means more of the 24 hours will be daylight and less night hours.

The Sun begins its orbit in the Fourth Co-ordinate in the East and orbits for 30 days. Continues to the Fifth Co-ordinate and orbits in the Fifth Coordinate 30 days. Then it goes to the Sixth Co-ordinate and orbits for 30 days. This completes the First Quarter. [April, May, June] The arrows on the left indicate the journey of the Sun during the First Quarter. Look on the diagram on page 152 with the beginning of the first quarter.

Beginning of the Sun's orbit

Figure 8.1

The beginning of the Sun's orbit with its rising in the East in the Fourth Co-ordinate. The Fourth Co-ordinate corresponds to the month of April. This marks the beginning of the Cosmic Year. It is

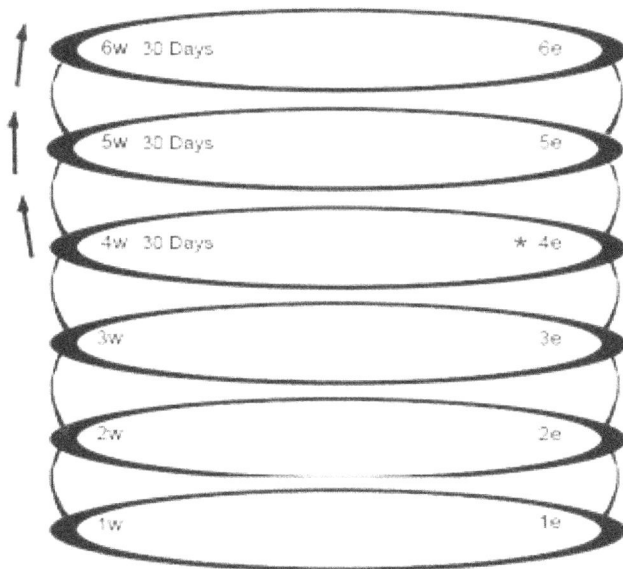

God who sets the point of the Sun's beginning. He had a Divine reason for giving the Children of Israel a new beginning of their year., a beginning that is identified with the beginning of the Cosmic Year. The Sun is seen as a star in the Fourth Co-ordinate East.

Naturally the laws of science can explain the motions of the heavens. In such case the stars are orbiting in a 360 degree circle. Since God created the laws of science, the initial orbits are determined by Him. But the Turning Back of the Sun twice in its yearly orbit is a phenomenon the scientists cannot explain. In this case, God intervened in the laws of science and the Sun obeyed. God always intervenes in nature for purposes known and unknown to us. Once the Sun completes its initial orbit, it will thus continue to orbit. The Divine command now becomes the Sun's natural orbit.

First Turning Back of the Sun

The Sun is going North and it turns back and begins to go South. Then at a certain point and time it turns back and begins to go North again. At certain times the Sun will be seen farther North and at other times will be seen farther South. It will be rather interesting to see how the Sun obeys this Divine command. As strange as it may sound, the Sun turns back immediately but not in an erratic manner. It must turn back in a manner friendly to the laws of science. We shall see that during the last day of the First Quarter, the Sun adjusts its south angle. And in the last day of the Third Quarter, it adjusts its angle North-ward. These two adjustments are responsible for the Turning Back of the Sun twice in its yearly orbit.

The South-east adjustment of the Sun's angle is fundamental in taking the Sun South but does not change the width and length of the daily circuit, except for slight variation. The volume of space of each circuit will not be completely independent. The circuit will overlap.

First Turning Back of the Sun

Figure 8.2

Circuit with adjusted South-East angle

Progress in South Direction

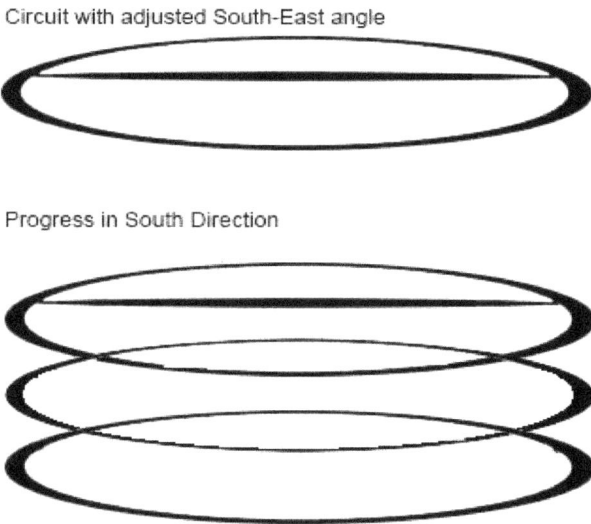

The Southeast adjusted angle is critical to the Turning Back of the Sun after the First Quarter. It means the Southeast angle becomes wider. This angle adjustment is responsible for the Sun's Turning Back the first time in its yearly orbit. The Sun turns back the last day of the First Quarter in the Sixth Co-ordinate. Rising in the East in the Sixth Co-ordinate, the Sun adjusts its angle South. At the completion of this circuit by going to the Sixth Co-ordinate in the West and returning to the Sixth Co-ordinate in the East, the angle adjustment is made and the Sun begins to go South. But it will orbit another 30 days in the Sixth Coordinate and then goes to the Fifth and Fourth Co-ordinates. This marks the end of the Second Quarter. The Sun will continue South to the First Co-ordinate. This will mark the Third Quarter. Notice the arrows on the left showing the beginning of the First Quarter, the Sun going North. The arrows on the right show the Sun going South.

First Turning Back of the Sun

Figure 8.3

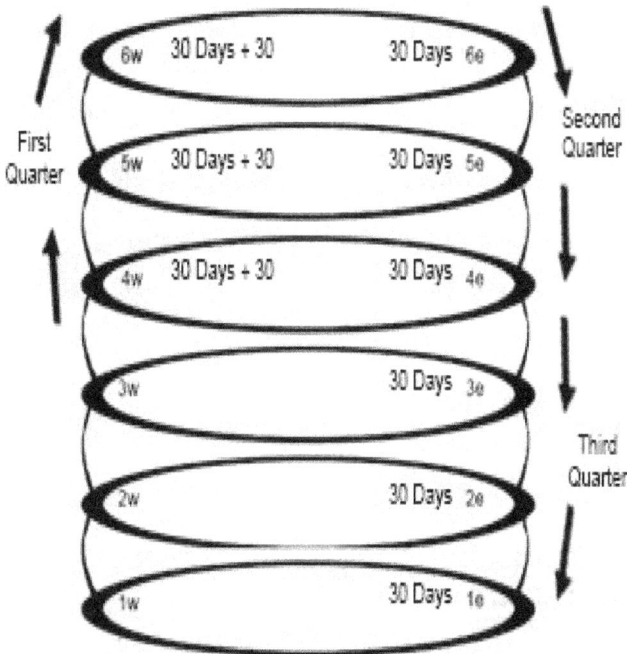

The arrows on the left indicate the first quarter of its orbit going north. The arrows on the right indicate its first Turning back going south . The Sun is now going South. It is now the month of December 31st.

The Second Turning Back of the Sun

The Sun must now turn back to go North. For this to happen the Sun must make a wider angle to the North when going from the West to the East in Co-ordinate 1.

Sun's North-western Angle Adjustment——Figure 8.4

This North-western angle adjustment occurs the last day in December, the end of the third quarter.

North Angle Adjustment

Progress In North Direction

As the Sun continues orbiting in Co-ordinate1, the wider Northern angle continues to take the Sun North. When it returns to Co-ordinate 3, it will have completed the Fourth Quarter and the first year.

Diagram showing the Turning Back of the Sun after the Third Quarter. Notice that Co-ordinates 4, 5, 6 remain intact, no orbital activities. The adjustment in The Sun's Northwestern angle is sufficient to turn it around north to begin its fourth quarter.

Figure 8.5

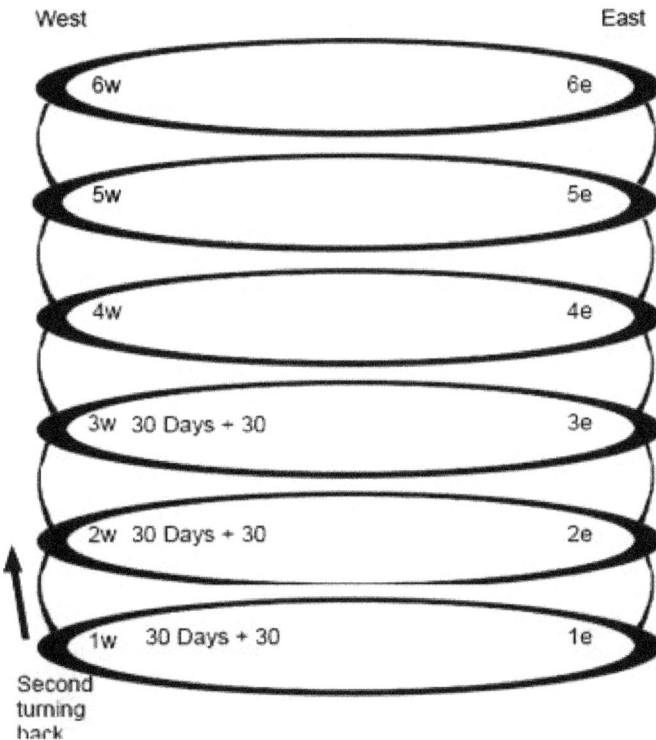

West   East

6w   6e

5w   5e

4w   4e

3w 30 Days + 30   3e

2w 30 Days + 30   2e

1w   30 Days + 30   1e

Second
turning
back

The Sun will orbit another 30 days in Co-ordinate 1, another 30 days orbiting in Co-ordinate 2, and another 30 days orbiting in Co-ordinate 3. This completes the Fourth Quarter. There are no orbital activities in Co-ordinates 4, 5, 6 because the Sun has not yet begun another yearly orbit.

The question naturally arises; do the laws of the Sun apply to other stars? It is quite possible but not all other stars. The Law of Variation would imply that the orbits of stars are not all one and the same. In a galaxy of 100 billion stars, orbits will differ. There are stars that orbit in a 360 degree circle which would be difficult to recognize their rising and setting. A Super Being would not be faced with this kind of difficulty. We humans need some planetary background to recognize their rising and setting. At an orbital velocity of 135 miles per second, the Sun travels 4,245,696,000 miles one year. A light year is 6,000,000,000,000 miles. At an orbital velocity of 135 miles per second, the Sun would take more than 1200 years to travel the <u>distance of a light year</u>.

Footnote: The book of Enoch the Prophet is one of the oldest religious books. After reading it several times, I wondered why it was not included in the Cannon of Scripture with the other 66 books which comprise the Bible. This feeling is not unique; many who have read the Book of Enoch have felt it should have been canonized. The Bible recognizes the nobility of Enoch's character and often quotes his writing extensively. "Enoch walked with God and was not for God took him" Genesis 5:24. Jude 1:6 "And the angels which kept not their own habitations, he hath reserved in everlasting chains under darkness unto the judgment of the great day". This quote is found in the Book of Enoch, Chap10:15, 16. Jude 1:14 "And Enoch also, the seventh from Adam, prophesied of these saying, Behold the Lord cometh with ten thousand of His saints, Enoch chap 2. Psalm 147:4 "He telleth the number of the stars; He called them all by their names."(Enoch Chap.XL111:1) The Sun begins its yearly orbit in the fourth coordinate Enoch Chap. LXX1:9,10. "In the same manner it goes forth in the first month by a great gate. It goes forth <u>through the fourth of the six gates, which are at the rising of the Sun"</u>

At an orbital velocity of 135 miles per second, the Sun travels 4,245,696,000 miles in one year. A light year is 6,000,000,000,000 miles, the distance light travels in one year. At an orbital velocity of 135 miles per second, the Sun would need more than 1200 years to travel the distance of a light year. The Sun needs more than 4,800 years to reach the nearest constellation, and more than 13,000 years to reach the constellation of Pro-cyan which is 11 light years away. The volume of space of the Sun's orbit is infinitesimal when compared to distances of light years. That is why positional changes in the orbits of these stars will not be observed. To us these stars remain static, stationary.

It must be pointed out that the 364 day calendar year herein computed is not based on the 360 degree premise of the Circle. The orbit of the Sun is very elongated. It is a semicircle. To be exact, it is a half circle but two halves make a whole. And by Turning Back twice in its orbit, duplicating it, it converts a semicircle into a complete circle. And that is why we cannot discover the profound wonders of the Universe by mere science. FAITH and REVELATIONS are critical factors. Every human being has a degree of faith, including scientists. The problem is that many have chosen to place their faith in things of the natural order, instead of God. "But God hath chosen the foolish things of the World to confound the wise; and God hath chosen the weak things of the World to confound the things that are mighty", 1 Corinthians 1:27. For it is written: "I will destroy the wisdom of the wise and will bring to nothing the understanding of the prudent", 1 Corinthians 1:19.

Having charted the Four Quarters of the Sun's yearly orbit, we now locate the volume of space of its actual orbit. The Sun is surrounded by 11 stars within a distance of 4.2 to 11 light years. The nearest of which is Proxima Centauri. Next is Bernard Star, a distance of 6 light years. The others are within a distance of 11 light years. Though the motions of these stars and the Sun's are continuous, the distance between them remain fundamentally unaltered. Any changes in distances are too insignificant to be mentioned. We can, therefore, ignore the motions of these stars and consider them stationary, and consider only the motions of the Sun. Within the volume of space created by these surrounding stars, the Sun has more than enough space for its yearly orbit.

Astronomy is one of the oldest sciences known to man. Indeed, new discoveries have been made in the heavens. But these discoveries have not implied nor suggested changes in the regular pattern in the heavens, such as changes in distances of the regular stars and constellations. Neither have these discoveries implied nor suggested additional stars and constellations within our neighborhood of stars. But precisely that would have been the case if the Sun was orbiting around the center of the Milky Way. And why aren't those scientists who say the Sun is orbiting around the center of the Milky Way not concerned about the Sun losing some of its planets, if not all of them, while traveling that lonely road around the center of the Galaxy? The point is that there is a part of man that would honor the immutability in nature, while at the same time that part would ignore and deny the omnipotence and supreme intelligence responsible for the immutability in nature.

The location of the Sun's Yearly Orbit.

Figure 8.6
Within the location of the eleven surrounding stars with distances from 4 to 11 light years, the Sun has more than enough space to complete its yearly orbit.

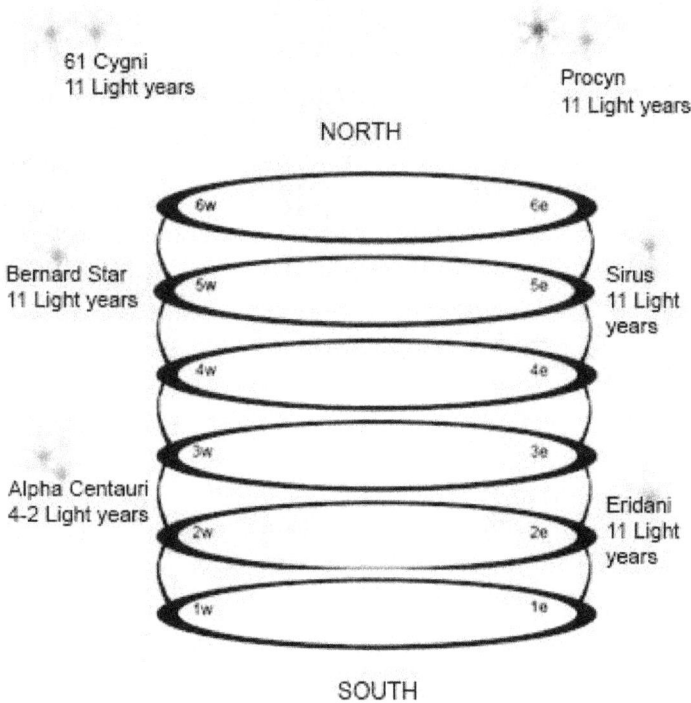

61 Cygni
11 Light years

Procyn
11 Light years

NORTH

Bernard Star
11 Light years

Sirus
11 Light years

Alpha Centauri
4-2 Light years

Eridani
11 Light years

SOUTH

Sun among 11 nearest stars

Figure 8.7

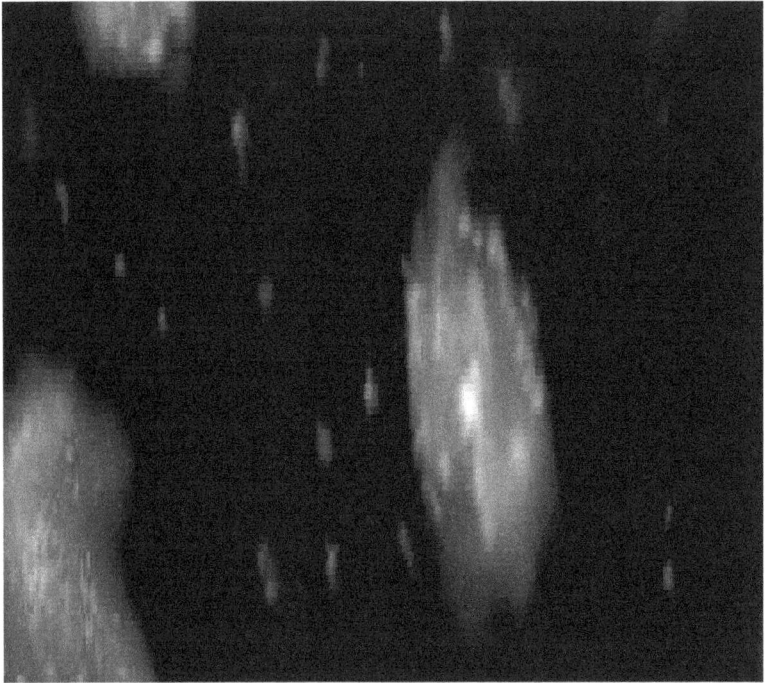

It is now obvious that the Sun is obeying a Divine command in turning back twice in its yearly orbit. Yet all the scientists in the World have failed to identify this phenomenon. They have said and will continue to say that the Sun is orbiting around the core of the Milky Way.

The great scientific discoveries of Sir Isaac Newton began from a point of simplicity. He used the known laws to explore the great unknown. One result was his articulation of the laws of Gravity and Motion. It simply began while he watched an apple fall from a tree to the ground. He reasoned that the force which caused it to fall to the ground could extend to the Moon and cause it to circle the Earth. He then applied this reasoning to the heavenly bodies. The force that causes the Moon to orbit the Earth could cause the obits of the celestial bodies.

Newton looked at a simple clock and watched the movement of its hands and reasoned that its slow sweeping hands were but the work of its internal gears. He applied this reasoning to the Cosmos. The motions of the Cosmos were but the work of internal gears, and coined the phrase, "The Clock Work Universe".

I must, therefore, reason that because the Sun turns back twice in its yearly orbit, its orbit is well restricted. I must further reason that because the Sun is surrounded by 11 other stars from a distance of 4 to 11 light years, it has more than enough space to complete its yearly orbit. Greater yet, the patterns of the heavens are regular and unchanging. The known constellation of stars can always be identified without any difficulty. Were the Sun orbiting around the core of the Milky Way, the patterns in the heavens would have been highly irregular. Joachim Ekruttis, therefore, is correct in saying that at times the Sun appears farther North and at other times it appears farther South.

As the orbit of the Sun is unique so is its purpose. It is the only star we know that critically relates to our World of 7 billion people. There is still much we do not know about our Universe. We do not know if there are other civilizations in the universe. We do not know if there are other systems in the Universe like our Solar System. Yet if we assume that there are, such assumptions would not make it a fact. We know there are three levels of intelligent existence: GOD, angels and man. God and Angels dwell in Heaven, the Universe above. While God rules in Heaven and the Universe below; angels have been given special duties relating to the Universe below.

To understand the phenomenon of the turning back of the Sun twice in its yearly orbit is to understand that God rules in nature. There could be no greater direct evidence than this.

Reconciling the Location of the Sun's Orbit with the Scientific Deduction in Chapter 6 that the Sun is a Member of the Three Star Constellation of Alpha Centauri

The location of the Sun's orbit has been identified for the first time in human history. It has not negated the fact that the Sun is the fourth member of the Alpha Centauri Star Constellation. Neither does it negate the fact that the Sun is also the twelfth member of the 11 closest stars. The reality is that in the Sun's yearly orbit of 364 circuits, it sustains the same distance relation with all the other members of the 3 star group. The only single exception to the location of the Sun's yearly orbit is that instead of orbiting a particular star of this group, it is orbiting according to a Divine command. This makes its orbit rather finite and restricted. It is this very nature of the Sun's orbit that gives us absolute certainty that our location of its orbit is correct.

This gives us an amazing scenario. We have the 12 windows of the Sun, the 12 months of the year, the 12 Tribes of Israel, and the 12 Disciples. Inevitably, this reveals the Divine workmanship in the fabric of the Universe. The physical and material reflect the spiritual —— AMAZING!! An unmistaken harmony in the Universe is obvious to all.

Infinite or Finite Universe: Open or Closed Door

I must now express my final thoughts on the model Universe. Is the Universe finite or infinite, open or closed door? God transcends time and space. The Universe cannot contain Him. He is eternal and infinite. By design he created the Universe to reflect His infinity. But the very nature of His infinity has sets a limit to the Universe. By design He had to decide how much of His infinity He wanted the Universe to reflect. So it reflects the measure he decreed before Creation. It is certainly the measure that brings glory to Him. So we can look to the heavens and ascribe to Him the glory of Creation. To us the Universe is infinite; we cannot explore its limits.

From the Divine perspective, we shall consider it a finite one, a closed door Universe. In this finite Universe, there are established boundaries known only to God. It is absolutely impossible for man to explore these boundaries. A description of a finite/ closed door Universe is one whose boundaries are encircled by cosmic clouds. Look again at the cause of cosmic clouds in chapter 3. The gravitational force of the Universe cannot contain the distribution of clouds perfectly. A perfect containment means that the distribution of clouds is balanced within the Universe and not be allowed to escape to its boundaries. The gravitational force of the stars at the outer reaches of the Universe would contain these clouds and also illuminate them.

By the design of the Universe as reflected in the structure of the atom and also by the spherical structure of bodies in space, these clouds would take a regular form. They would encircle the Universe in regular density and volume. However, it would be difficult, if not impossible, for any scientific instrument to discover the limits of the Universe for the fundamental reason that the inner part of this Cosmic Circle of clouds would be illuminated not only by the stars at the outer reaches of the Universe, but by others as well. The end result would be that from whatever direction a scientific instrument was pointed, it would have only detected luminous clouds. And owing to distance of light years away, many scientists would interpret them to be other galaxies

Equally true, the outer part of this Cosmic Circle of clouds would be infinitely black. Indeed, it would fit the Biblical description, **outer darkness.** It is as though the Creator leans His back against this wall, not to take a rest, but to challenge anyone to go beyond. So, the great and eternal secrets of life, His omnipotence, and His eternal existence are only fully known to Him. There are those that would question the concept of a Closed Door Universe, or finite Universe, one with established boundaries and choose instead an Open Door Universe, one with no established boundaries, forgetting that the latter is contrary to the laws of nature, cosmic or biological. Every single law of nature carries along with it a kind of inevitability, which is its imposing limitation. So, a star orbits within the limits of its own orbit. A human being grows to his own biological limit. If our Universe is the Closed Door Model, the description here given is adequate.

Model Universe

Figure 8..8

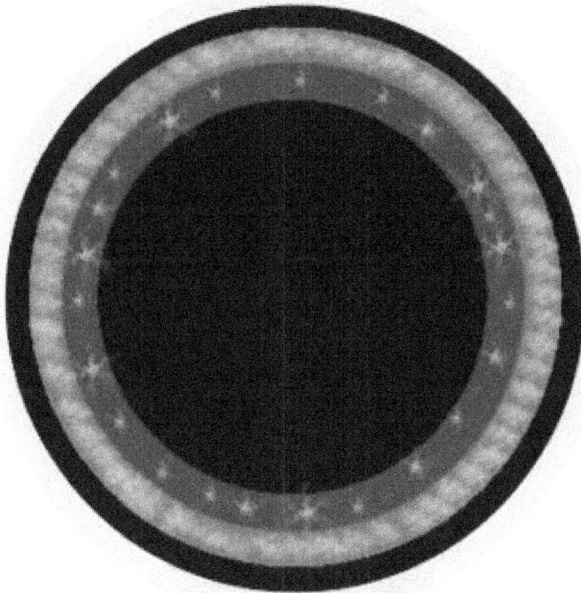

Circle 1 represents stars at the outer reaches of the Universe
Circle 2 represents luminous clouds at the boundaries of the Universe
Circle 3 represents outer darkness, the other side of the Cosmic
Clouds. Circle 4 represents infinity – Divinity.

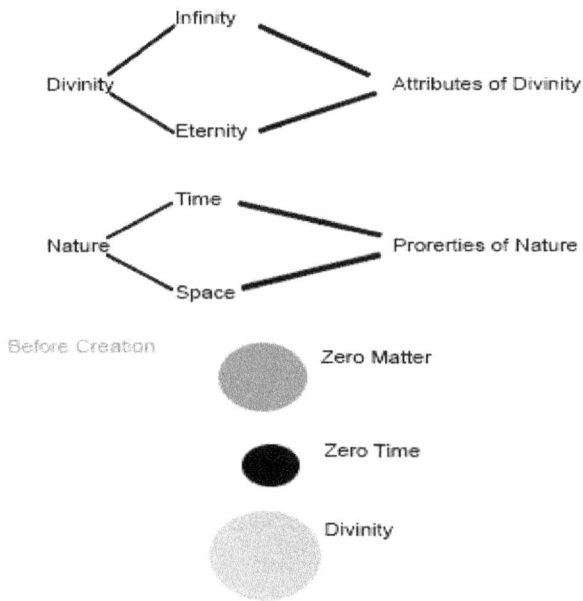

Before Creation there was no matter or any other reality except Divinity. Since there was no other reality that could limit Divinity, Divinity must, therefore, be infinite. Divinity can only be limited by ITSELF. But with respect to Divinity, limitation only has to do with manifestation and not a change in nature. God can and does limit His manifestation, but He cannot change His nature. A change in nature means He would cease to be God. This takes us to the IMMUTABILITY of God.

The Immutability of God means that He cannot change any of His attributes even if He would like to. God's existence depends not on His will, but on His nature. It means His immutability is rooted in His eternal existence, an existence without beginning. Since God did not create Himself, He cannot change His nature or anything in His nature. Therefore, God's immutability is rooted in His eternal nature. We cannot fully comprehend God's eternal nature. Only God Himself fully knows and understands.

With respect to God's infinity, it must be emphasized it is not an outgrowth from a finite state or a development in Divinity. It must also be emphasized that the laws of nature point to immutability of God's nature. Since there is a God, He must be eternal, immutable and infinite in all His attributes. Some of His attributes are: love, holiness, power, wisdom, justice and mercy. With respect to space, He is not limited to space. The Universe cannot contain Him.

Chapter 1: The Underlying Theological Postulate

Any scientific description of the Universe from a Theological Perspective must have a correct underlying Theological Postulate which does not attempt to prove the existence of God. Rather; it assumes God's existence and asserts that belief in His existence is compatible with man's intellectual endowment, moral, and spiritual nature. It recognizes the revelations of God attributes as the inherent cause of His existence for the fundamental reason that He exists by reason of His nature, and not by His Will. Therefore, He must exist from all eternity.

The Postulate does not attempt to prove God's existence because it would not be possible to do so. One who owes his existence to a known point in time could not prove the existence of an Eternal Being. Neither could one who is limited to time and space prove the existence of a Being who transcends time and space. One could not explain God's attributes except by the evidence seen in nature. Most definitely, the evidence of omnipotence, omniscience, and omnipresence are seen in nature. These evidences are not only logical, but also irrefutable. What this means is that God is the First Source of knowledge. Thus, He is the indispensable factor in all knowledge. In this regard, science only makes sense when God is acknowledged, implied or expressed. Without the Divine source of knowledge, man's moral and spiritual darkness would have been greater than the darkness of outer space. In short, knowledge is a priceless gift from God and the Holy Bible is the greatest recorded source of knowledge. Those who refuse to recognize this fact will live in intellectual, moral, and spiritual darkness.

Since the Holy Bible is the fount of knowledge, we should carefully examine it to see what it says on the most fundamental subjects to our understanding of the Universe. A thorough understanding of the

Universe will lead us to its cause, purpose and destiny. For indeed, the concepts of the Theologian are derived from the Bible, though may be developed from the information of modern science. Herein is addressed, the Bible itself, God Himself, and the Works of God——Creation.

The Bible

The Bible is not just an ordinary book written by man, but is the holy, inspired, and revealed information of God. Man could have known there is a God without reading the Bible because God has revealed Himself in man's conscience . But man can ignore the just pleadings of his conscience. However, man could not correctly articulate the nature and attributes of God without the revelations in the Bible.

The Bible is the text book of the Christian Religion. It is undoubtedly the best known book, the best read, and the oldest text book. Its universality is accompanied by an inescapable justification. There is a reason for its universality. The reason is its Divine Authorship. A book expresses the mind, the thoughts of its author. The Bible in its essential nature expresses the mind of God regarding His Purpose in Creation, the Fall of man, Redemption, and the Destiny of Mankind. And how could we know the mind of God except by Divine revelations.

The predictive aspect of the Bible also supports its claim of Divine Authorship. Many predictions of future events were made, most of which have already been fulfilled. Some are yet to be fulfilled. The prophets who made these predictions did not claim they were the results of their natural insight, but were due to instantaneous endowment of supernatural knowledge. The claim of Divine Authorship does not mean that most of the information of the Bible is comprised of revelations or predictions. It means the most fundamental information was derived by Divine revelations. Man could not have known about the attributes of God except by Divine revelation, neither about Creation, the Fall and the Destiny of Mankind.

The Bible, for the most part, is a record of God's relationship with mankind. In light of this, the Bible is naturally cumbered with secular details, man's behavior, and God's response to that behavior. Much of the Bible expresses man's despair as well as his courage, faith and determination to survive. Much of the Bible clearly mirrors man's true nature, so that one can predict how man, under given circumstances, will act and react. But the claim of Divine Authorship is also supported by the indestructibility of the Bible, its character and its unity. Though the Old Testament Scriptures were officially accepted as the revealed and inspired Word of God by the Church Council A.D. 200 and the New Testament Scriptures A.D. 400, the Bible is dated back to a much earlier period to God's first dealings with man.

The Postulate with Respect to God

A reasonable understanding of our Universe hinges on the realization that God is the indispensable factor in all of knowledge. Yet, the diverse views and ideas of God would naturally belie the truth of this statement. Here, one would be forced upon a most precarious path, fraught with great uncertainties. And religions do not help to alleviate the pains inflicted on the journey of this path, because religion must first answer the question, so frequently asked among them, Which is the right religion? While the major religions share the burden of declaring the underlying message of the LIFE BEYOND, their ideas of God make them fundamentally different. Even in the Christian Religion all are not agreed on the Doctrine of the Trinity. From our point of view, it is important that we here and now articulate the nature and attributes of God. This is critical to a true scientific description of the Universe. The only authentic source of this knowledge is the Bible. The Bible clearly teaches:

1) The Eternity and Self Existence of God. By the eternity of God is meant that He is without beginning, that there was never a time when God was not existing. He always was, is and shall be. We do not pretend to understand the eternity of God beyond this level. We are satisfied that God exists and because He exists He must be eternal. By the self-existence of God is also meant that His existence and nature are independent of matter as we know it. As Eternal, God transcends time. The Bible does not argue or attempt to prove His existence. It has assumed His existence, that mankind believes there is a God, and that this belief is innate in man. You see, when the Bible was written there were no doctrines of evolution nor the Big Bang Theory.

2) The Infinity of God. By the infinity of God is meant that He is free from all limitations. In the first place, He is not bound by time. In the second place, He is not bound by space. He transcends space and time. In the third place, He is not limited in any of His attributes (spiritual infinity). An Infinite God can limit His manifestation. This is quite often the case. Infinity in His attributes means that He is omnipotent (all powerful), omnipresent (everywhere present), omniscient (all knowledge and wisdom); and His love, justice, mercy, holiness are without limits.

3) The Spiritual Nature of God. By the spiritual nature of God is meant that He is a spiritual substance. He is not made of matter as we know it. Due to His spiritual nature He is immortal, immutable and invisible. Being immortal He can never cease to exist. Being immutable, His attributes cannot be changed or improved one way or

another. Being invisible He cannot be seen by the human eye. He can, however, manifest Himself in a form that the human eye can recognize. The eternity and self-existence of God, His infinity, His spiritual nature— all imply that He is the LIVING GOD, that He had the ability to create the Universe and He did.

4).The character of God. By the character of God is meant the sum total of all His Divine activities and actions. All Divine activities and actions must be rooted in His holy nature. Herein lies the foundation of the moral Universe. The God of time, space, and nature cannot do anything that is wrong. Because man was created in the His image, he is always inclined to do that which is right, even though he often does that which is wrong.

In our moral Universe, the actions of men are always questioned as to whether they are wrong or right. Their actions of wrong or right determine their character. To us, therefore, character is of utmost importance. The Bible contains a number of Divine actions which seem to have contradicted our definition of the character of God. While we cannot enumerate all of them, we wish to mention some of the most obvious and fundamental ones in relation to mankind. We here mention the following ones:

1) SIN. God is not the Creator or the Cause of sin. Sin resulted from the Free Choice of man in rebellion to a known law or a Divine command. Man's Free Choice is the faculty of his soul that elevates him above the level of the animal creation and makes him like his Creator. The Free Choice with which man was created must be considered a holy act on the part of God.

2) The Destruction of the Antediluvian World. The physical destruction of mankind, except for 8 persons was a judicial act of God, simply because mankind had consistently revolted from Him in sinning beyond the **tolerable limit**. God has a sovereign right to rule

the moral Universe, but can only do so in keeping with His own moral nature. His moral nature dictated the destruction of the Antediluvian World, and yet their preservation by saving 8 persons. The surprise is not really about the destruction of the Antediluvian World. It is really the tolerance of God's holy nature of sin to the degree that He permitted them to sin for years.

One may form the same conclusion of some other Divine judicial acts such as the Destruction of Sodom and Gomorrah, the Destruction of the Egyptians in the Red Sea, and the Wars of Israel with the nations of Canaan. All these must be seen as Divine acts of a punitive nature, executed in order to preserve the human family, in keeping with the Divine promise of Redemption. We must understand that the holy nature of God cannot tolerate sin beyond a certain limit. One can also see that the Divine choice of the nation of Israel was for the purpose of a specific role in the plan of Redemption, and was not an exercise of mere preference of one nation above the others.

Our conviction remains rooted in the truth that God's character is the foundation of our moral Universe and that foundation is firm and steadfast. Today men act and react in view of moral ends. And if such actions and reactions are wrong, by their faculty of conscience, they feel self-condemned. Indeed, a sense of guilty conscience causes them great human pain. The knowledge of the self-revealed personal God of the Bible is indispensable to our knowledge of the Universe

Finally, the underlying Theological Postulate shows the Universe as the works of God. In this regard, the two great contradistinctions in the formation of the Universe have been established: the Miraculous Aspect and the Scientific Aspect. In the Miraculous Aspect, God created the conditions of matter – space and heat. This cannot be explained by the laws of science. Subsequent to the creation of these conditions, the formation of the Universe assumed the laws of science. From here on, to a point, one could follow the development in the process of the formation of the Universe.

Also important in the Scientific Aspect is the premise: every phenomenon in nature, event, occurrence, or condition is the result of some other phenomenon, event, occurrence or condition in nature. This was the basis for the scientific deductions in this book.

Chapters 2 and 3: Scientific Theories: Review and Response

A theory summarizes the scientific information of a particular subject. Reviewing and responding to some of the better known theories is in keeping with the objective of this book of reconciling science and Theology, in the hope of providing a better understanding of our Universe. It is of great significance that a Theological Theory of the formation of the Universe, the Theo-cosmos Theory has been herein formulated. Such theories as the Development of the Earth, the Hot Big Bang, its Inflationary Version, the Evolution of the Solar System, and the Evolution of Life are reviewed and adequate responses given. Other theories such as the Expanding Universe, Newton Mechanics, Relativity and Quantum Mechanics have been reviewed.

The Origin of the Earth

Scientists believe that the Solar System was once a cloud of dust and gas. They believe that this cloud of dust and gas was the result of a star explosion 5 billion years ago. They claim the Sun is a third generation star. According to them, an earlier generation of stars were necessary to create the heavier element of the present generation of stars.

G.P. Kuiper states the Theory best. According to him, the Sun was formed by the contraction of a rotating cloud of gas. Some of the matter escaped the process of formation of the Sun. While the Sun was forming, the matter that escaped the formation process broke up into smaller volumes and formed into nine planets. Kuiper assumes that the Sun and the planets were formed simultaneously and that when the Sun started to rotate, large portions of its outer masses were blown into space and formed into planets. Some of their masses were also lost to evaporation. <u>Atlas of the Universe by Thomas Nelson and Sons,1961 Ed, p102.</u>

Our response to this Theory is similar to our response to the Big Bang Theory with regards to the duration of the process of formation as the basis for calculating the age of the Solar System. As we have stated in the THEO-COSMOS THEORY that the Universe was once in a state of dust and gas in its early formation, we accept that part of the theory to be authentic. While we accept that parts of the Universe are older than other parts, we refuse to accept that the age gap between any two parts could be 5 to 10 billion years, or that the Sun is a second or third generation star. We take the position that the Solar System was formed in the beginning with the rest of the Universe from gas of clouds and dust; but the conditions for life and life forms were not created on the Earth until sometime after.

The Development of the Earth

The development of the Earth is interestingly explained by the geologists. They explain that the present form and condition of the Earth took a period of approximately 5 billion years. They divide this time into six Geological Periods. We do not know if these periods were intended to correspond with the 6 Creative Days of Genesis chapter 1.

The First Geological Era——Archeozoic. This era lasted 1.5 billion years. During this period the Earth began its existence as a body of hot liquid and gas. As the hot liquid cooled, it formed into solid rocks. In the process a thin crust began forming into the Earth's surface. This crust often broke as more liquid rock spilled on. The weight of the liquid rock caused the crust to sink in some areas and to rise in other areas. While the Earth continued cooling, condensation in the Atmosphere began in the form of rain. It rained for millions of years until the Earth became cool enough, but the Earth's center remained a hot dense liquid.

After the Earth's crust was formed, parts of it were broken up by thousands of years of rains and winds. Rocks were broken into small particles and carried thousands of square miles on the Earth's surface. This era ended with the beginning of life.

The Second Geological Era—— Proterozoic. This era lasted 1.5 billion years. It saw the development of plant and animal life in the oceans. Later animals began to live on land. Further changes in the Earth's surface began to occur through volcanic activities, causing parts of the crust to merge while other parts plunged beneath the seas. Many mineral deposits were formed by the cooling process of the lava.

The Third Geological Era——Paleozoic. This era lasted 800 million years. A major feature of this era was that plant and animal life became more complicated and abundant. Sea animals began moving out of the waters to live on land. Many types of insects developed. Great coal fields formed from buried plants and trees. Further changes in the Earth's crust occurred. Great layers of ice formed at the North and South Poles. Winds and rains effected further changes in the Earth's surface.

The Forth Geological Era —— Mesozoic. This era lasted about 800 million years. Remarkable biological changes took places within many animals. The reptiles became the dominant animals of this period Noted also was the appearance of fruit trees.

The Fifth Geological Era—— Cenozoic. This period lasted about 399 million years. Very early, warm blooded animals replaced the reptiles. Dominant among them were the mammals. Elephants and horses were among the first to appear. In the meantime, changes continued to occur in the Earth's surface.

The Six Geological Era. This was the shortest era. It began about a million years ago. At the beginning man appeared on the scene but not in his present form. Noticeable in this period was the covering of the Earth's surface with ice. The Earth is said to have been covered with ice from the North to the South Pole four times.

The scientific theories of the origin of the Earth and its development are not theories we can completely reject. Without hesitation, we can emphatically reject the accidental aspects of the theories with respect to the origin of the Earth and the evolution of life. But we have no Biblical ground to reject the age of the Earth and its stages of development. We would question the durations in the stages of development in the plant and animal kingdoms. But we cannot condemn or reject the stages in the Earth's development as explained by the geologists since the Biblical Account begins with an Earth already formed except for the conditions of life and life forms. The Bible does not tell us how God formed the Earth; it only gives some details of the creation of life conditions and life forms.

The Big Bang Theory

The explosion of the Cosmic Atom 100 million miles in diameter marked the birth of the Universe. About 20 billion years ago, this super-large atom of protons, neutrons, and electrons exploded scattering its mass into space. Minutes after the explosion, temperature exceeded several billion degrees. Particles began to form nuclei. About 30 million years latter temperature had fallen to a few thousand degrees and some of the gas contracted into dust. The force of Gravity began to attract the dust into great masses. From these masses the galaxies of stars originated. This Theory is attributed to French Scientist, Lemaitre (Julien de La Mettrie).

ANOTHER VERSION: The Big Bang occurred about 15 billion years ago. The entire Universe exploded out of a point of infinite density. The explosion is uniform and featureless.

Inflationary Expansion. The Universe expands dramatically. Matter in the Universe is still evenly distributed.

300,000 YEARS AFTER THE BIG BANG. Huge clouds of matter, 500 million light years in length and larger began to condense.

Galaxy Formation. The first galaxies appear about 200 million years after the Big Bang.

TODAY. Fifteen billion years after the Big Bang, stars and galaxies have evolved out of the clouds of matter. The lights from events that occurred near the start of the Universe have traveled for about 15 billion years to reach the Earth.

This most recent version of the Hot Big Bang Theory has left room for wild speculations as to the origin of space. It has not given a definitive statement about space. Space would have to be a prerequisite of that Primeval Atom of 100 million miles in diameter. The second flaw of the theory is that it has said nothing as to how the matter of this Primeval Atom originated. For these two fundamental reasons, there is not much substance to the Theory. And by virtue of these inherent deficiencies, the Theory has not given us the cosmological secrets of the Universe.

As we have quickly pointed out the flaws of the Theory and without hesitation rejected them, we must concede the favorable features. We see in the favorable features of the Theory the equivalent of a miracle. For matter that volume and infinite density to have exploded could only be a miracle. The scientists have unconsciously admitted this. We know that the splitting of the atom by man took diligent and systematic efforts over a long period of time. The explosion of the Primeval Atom implies external intervention. This atom when compared to the size of the Universe could only be regarded as less than nothing. The development in the evolution of the Universe could also be regarded as a miracle. So in essence, the Theory shows how something was made out of nothing.

We take this position of reconciliation because to utterly reject the Theory would be placing severe limitation on an Infinite and

Omnipotent Creator. Yet we cannot say with any degree of certainty there was a Big Bang. However, to say that God could not create a mass of matter of infinite density and then exploded it in forming the Universe is, indeed, the placing of limitation on the Omnipotent Creator. But I am fully convinced that by correctly interpreting the scientific references in the Bible of Creation and of the nature of God in conjunction with the correct interpretation of the laws of science, we can perceive exactly how God formed the Universe.

If one believes the whole visible Universe came into being instantly, in its present form, then there is nothing else one can know about its formation. Many Christians and Theologians share this belief. Dr. Thiessen author of Introductory Lectures in Systematic Theology shares this belief. And I suspect many other outstanding theologians share this belief. I now quote Dr. Thiessen: "By immediate Creation, we mean that free act of the Triune God whereby in the beginning and for His own glory, without the use of preexisting materials or secondary causes, He brought into being, immediately and instantaneously, the whole visible and invisible Universe.

The Theo-Cosmos Theory of the Formation of the Universe

There are two great contradistinctions in this Theory: the Miraculous Aspect and the Scientific Aspect

The Miraculous Aspect. This aspect in the formation of the Universe had to do with the creation of the conditions of matter: space, heat, and cold. Empty space was first to be created. Before the Creation of the Universe The Eternal God existed in the fullness of all His attributes. One of His attributes is an infinity of majestic Light. In consideration of Creation, He limited the manifestation of that majesty. The immediate result was space. An Infinite Being can limit His manifestation.

The Scientific Aspect. Following the creation of space, God then commanded the other conditions of matter into existence. The whole spatial Universe was subjected to intense heat — matter in the form of hot gas. From this point, the Universe began to form according to the laws of science. Most of the hot gas cooled and contracted into **cold invisible matter.** The rest cooled and contracted into clouds of dust above. The **cold invisible matter** far exceeds the clouds of dust. The clouds of dust then floated on. With the far greater amount of **cold invisible matter,** the condition of weightlessness was created — the foundation of the Universe was established. Many Science Books have not mentioned the cause of the weightlessness of outer space.

Apparently it seemed unlikely that the greater amount of hot gas should have cooled into **cold invisible matter** and the less amount into clouds of dust. But why when a cup of hot milk cools, the cream is thicker than the rest of it? Less than one percent of that cup of milk formed into cream. We could think of the clouds of dust as the Cosmic Cream at the beginning of the Universe.

This formative stage of the Universe is clearly implied in Job 38:37, 38: "Who can number the clouds (stars) in wisdom? Or can stay the bottles of heaven, when the dust groweth into hardness and the clods cleave fast together?" Both the early formative state of the Universe and the theory of star formation are here implied. If you had read the entire chapter you would have observed that the Almighty was giving Job a science lecture in which Astronomy and Cosmology were among the main points. Notice that in verses 31 and 32 the emphasis was on Astronomy. The Almighty lectured Job on four constellations of stars. "Canst thou bind the sweet influences of **Plades**, or loose the bands of ORION? Canst thou bring forth **Mazzaroth** in his season? Or canst thou guide **Arcturus** with his sons?" Then in verses 37 and 38 the Almighty was plainly saying to Job, "Did you know that all the stars of heaven were once in a state of dust but at My command they were formed into stars of light?"

The Expanding Universe

The Big Bang Theory is closely connected with the expanding Universe. The rate of expansion, immediately, after the explosion was incredibly great. With the passing of time of billions of years, the rate of expansion greatly reduced. Today the Universe continues to expand but just at the critical rate to balance the attraction of Gravity.

The Expanding Universe is the subject of the third chapter in the book, A Brief History of Time written by Stephen Hawking a theoretical physicist. In the third chapter he emphasizes the spatial expansion of the Universe. The distances between galaxies are increasing constantly. The Hubble Discovery in the 1920's showed that the galaxies at the outer reaches of the Universe were moving away and the farther away, the faster they were moving. This discovery forms the foundation of Stephen Hawking's argument for the Expanding Universe. He argues three points to support the Theory of the Expanding Universe. They are the Doppler Effect, the Common Similarity with parts of the Universe, and the Microwaves Radiation.

The Doppler Effect is applicable to sound as well as radiation. In the case of sound, the frequency is higher moving towards its source, depending on the speed at which one moves. The frequency or sound is lower, moving away from the source, depending also on the speed at which one moves. In the case of radiation, measuring the distance of a star, a shift to the blue end of the spectrum means that the star is moving towards the measuring instrument. A shift to the red end means that the star is moving away. Accordingly, all the galaxies in Hubble's discovery showed a red shift: the galaxies were moving away, meaning the Universe is expanding.

Response The discovery of a new phenomenon of nature does not necessarily mean the discovery of a new law of nature. But it does call for a closer examination of the known laws. Indeed, it is this closer examination that will determine whether a new law has been discovered. Until that is done, we should expect that the already known laws do apply. Thus we expect that the laws that govern the galaxy of the Milky Way also govern the galaxies of the Hubble's discovery. Our examination of these known laws and their application must be thorough; we must search diligently.

Our concept of the center of the Universe is that region that has the largest number of galaxies when compared with any other region, or combinations of regions. The greater configuration of mass, the greater the slowing down effect of the Centripetal Force: galaxies at the center move slower. The farther away a galaxy is from the center, the faster is its speed. Velocities of motions do not contradict the known laws of science. Thus the faster motions of galaxies at the outer reaches of the Universe are not evidences of an expanding Universe. An evidence of an expanding Universe would be an increase in distances among the stars in the Milky Way and noticeable formation of new stars between the expanded spaces. Only a substantial expansion, one in which both volume of matter and distances increase could be considered an expansion. The scientists have not informed us of such an expansion.

It cannot be denied that the vantage point from which the Universe is viewed must be considered. Looking from the outer reaches of the Universe at its center, one would see the galaxies spreading apart, though they are not thus seen by scientists. God may have created an infinite Universe in the initial act of Creation in the Dateless Past. Though it would have involved an indefinite period of time, it would not necessarily have an expansion aspect. This point can be illustrated by the analogy of a building and its builder. A very large building may have been completed in a non-stopped construction. On the other hand, it may have been completed by additional constructions at different times, like a poor man who first built a two bedroom house and later built additional rooms. In this case we have favored the former as the way God constructed the Universe. If there were any additional expansions, as the latter, that expansion ended the day God created the life forms and conditions on Earth.

The Theory of the Evolution of Life

In their Book Science Matter, Hazen and Trefil best sum up the Theory. After stating the renowned premise of the Theory that all life forms evolved from earlier and simpler forms, they argue a two- step process: Chemical and the Biological Evolution. The Chemical Evolution was critical to the Biological Evolution and so naturally preceded it. In the Chemical Evolution of life, the oceans became indispensable. They were the mixing bowl. The Earth's early Atmosphere played an equally important role in the Chemical Evolution. It provided such elements as carbon, hydrogen, oxygen, phosphorus and sulfur. These gases mixed with the surface layers of the oceans. This mixture formed the complex molecules necessary for life. The process involved millions of years.

Once the Chemical Evolution occurred, the Biological Evolution began. Unlike the Chemical Evolution which had taken millions of years to produce the first living cell, the Biological Evolution took a brief time in comparison. In a short period of time that first living cell multiplied itself to the degree that it filled the oceans. The Theory concludes that the diversities of all life forms came from this first living cell. However, they have reached no definite conclusions as to how and when the first living cell originated. "We do not know how life arose. This remains the greatest gap in our knowledge." Science Matters op cit. p.247

Hazen and Trefil argue that during 1953 Stanley Miller and Harnold Urey at the University of Chicago conducted an experiment to determine what natural process could produce the complex molecules necessary for life. The results of the experiment revealed amino acid is the building block for proteins. The longer the experiment lasted, the more diverse and concentrated the matter in the experiment became. It was to the degree that some people had thought new and dangerous forms of life might emerge from the test tube.

Response

Some evolutionists misinterpret Darwin's Theory of Natural Selection to explain the diversities in the animal kingdom. The focus of the Theory is on the survival of a species. The Evolution of Life Theory is unable to answer at least three questions of life. First, it has failed to explain exactly when and how life began. Expressing an idea that the first living cell may have taken millions of years to form is not really telling us when life originated. And identifying the chemical elements of life is not really telling us how life was formed or evolved. The fact that the evolutionists are unable to put the chemical elements together to create life is the evidence that the Theory has not explained **when** and how life began.

Second, the Theory cannot account for the millions of species in the animal kingdom. If life evolved from a single living cell, how came these species of such great diversities? But the true meaning of Natural Selection lies buried in certain differences within a species. Looking at the Doctrine of Natural Selection for a brief moment, it will be obvious that Natural Selection perpetuates a species and says nothing about how a species originated. Look at the premise in the words of Darwin: "The preservation of favorable individual differences and variations, and the destruction of those which are injurious, I have called Natural Selection, or the Survival of the Fittest." The Origin of the Species by Charles Darwin, University of Oxford Press, Sixth Ed.P181. It is clear that Natural Selection deals with the preservation of species and not the origin .

Third, the Theory is completely silent on the most vital question of life, the question of intelligence. In the plant kingdom the characteristic of intelligence is absent. This is what to be expected from an accidental cause. Yet, the greatest problem with the Theory is its failure to explain the millions of species in the animal kingdom and the moral and spiritual code of human nature.

Newton Mechanics

The Theories of Gravity and Motion otherwise called Classic Mechanics were formulated by the brilliant English scientist, mathematician, physicist, and astronomer, Sir Isaac Newton (1642–1727). Newton's theories were first published in 1687. He was preoccupied with the motions in the Universe, particularly those of celestial bodies. He saw that the motions of these bodies were regular and systematic. He had been considering motions on Earth too. He knew that motions were not spontaneous. To get a stone moving would require an external force. The larger the stone, the greater would be the force required to get it moving. There was always a tendency of resistance to a change of position. A stone rolling down a hill will continue rolling unless it is stopped by an external force. This tendency to resist a change in position, Newton calls "inertia" Newton found that all masses have a degree of inertia, which can be overcome. To overcome the inertia of a mass, means a change in position of that mass. To Newton there were only two basic kinds of motion: uniform motion and accelerated motions. Uniform motion is continuous motion without any change in speed. Accelerated motion is a motion in which there is a change of speed.

Newton sums up the motions of the Universe in the following three laws:

1) A body at rest will remain at rest, and a body in motion will continue to move in a straight line at the same speed, unless the body is acted upon by an outside force.

2) Acceleration, or change in velocity is directly proportional to the force and inversely proportional to the mass. It is evident that this law deals primarily with motions on Earth since celestial bodies are in a uniform state of motion. The motions of motor vehicles best illustrate this law. The greater the force is the greater the acceleration, direct relation. The greater the mass the slower the acceleration. A heavy motor vehicle will move off more slowly than a light one with equal force, inverse relation.

3) Every action causes an equal and opposite reaction. When you push against an object that object pushes against you with equal force. The force of the push is always balanced. This principle applies equally to motions on Earth as well as celestial motions. The gravitation force exerted by the Sun on the planets is in turn exerted by the planets on the Sun. Because of the greater mass of the Sun (98 percent of the matter in the Solar System) the planets orbit the Sun.

Gravity

As an astronomer Newton knew that the heavenly bodies move in circular line. He reasoned that there was some force that causes this to happen. He called this force Gravity. Newton's concept of Gravity was conceived when he saw an apple fall from a tree while he was able to see the Moon. He reasoned that the Earth's gravity which forced the apple to the ground could have extended to the Moon and caused it to circle the Earth instead of falling to the Earth. On this premise, he articulated three laws of Gravity known as The Universal Law of Gravitation:

1) Everybody attracts every other body in the Universe. This law implies that a body will orbit the closest center of gravity.

2) The amount of attraction is directly proportional to mass. The size of the mass determines its gravitational force, or which of the masses in a group is the center of gravity.

3) The amount of attraction is inversely proportional to distance. The farther the distance between two bodies, the less the force of gravity is felt, the one body on the other. Because of the relatively closer distance of the Moon (240,000 miles) than the Sun (93,000,000 miles), the Moon's gravity is stronger on the ocean's tides than the Sun's

RESPONSE: Newton Mechanics is not a perfect Theory for the reasons that he did not describe the nature nor a medium through which Gravity works. Some of his fellow scientists recognized these inadequacies and brought them to his attention.t. They suggested some microphysical entities and processes, but did not explain what these entities and processes were.

Sir Isaac Newton did not explain that the force of Gravity was magnetic attracting force. He did not explain exactly how Gravity works over long distances, and never seemed to have had the explanation. "To us it is enough that gravity does really exist, and act according to the laws which have been explained, and abundantly serves to account for all motions of celestial bodies and our seas." Encyclopedia of Philosophy by Cromwell and Mac Milan Inc., 1967 Ed. V1. 5. P 490.

However, Gravity is an in-escapable reality of nature, permeating all physical phenomena as well as the simple everyday things of Earth.

The Theory of Quantum Mechanics

The Theory of Quantum Mechanics is one of the many theories developed as a result of the study of radiation of light. It was proposed in the 1920's by German Scientist Werner Heisenberg (1901-1976). Like most theories, it has practical importance in understanding the nature of matter and in interpreting and understanding some laws of nature. Theories of the radiation of light led to the discovery of the speed of light (186,000 miles per second) between 1676 and 1865, and the concept of the Black Hole 1783.

Astronomers have made extensive studies of the radiation of light and thereby able to determine the chemical elements of the stars, their luminosities, temperatures, and distances. In truth and in fact, the study of the radiation of light is the foundation of Astronomy. The microcosm of Quantum Mechanics was first seen in the Theory of Determinism proposed by French scientist Laplace, at the beginning of the Nineteenth Century. Laplace is reported to have said that he swept the heavens with his telescope but could not find God. It was, therefore, not unnatural that he proposed the Theory of Determinism. The substance of Determinism is that a Supreme Intelligence possessing a knowledge of the Newtonian Laws of nature and a knowledge of the positions and velocities of all particles in the Universe, at any moment could determine the state of the Universe at any other time.

Without hesitation, many other scientists attacked Determinism. In their endeavor, a number of theories were developed. Quantum Mechanics was one of those theories blossomed and matured into fruits; it is a Theory recognized by many scientists as one the great theories of the Twentieth Century. While the microcosm of Quantum Mechanics could be clearly seen in the Theory of Determinism in that it was one of those theories developed as a result, Quantum Mechanics was a direct offspring of Quantum Theory proposed by the German scientist Max Plank in the early Nineteenth Century.

Plank's Quantum Theory is significant because it has defined for us the Quantum World to the degree that it lets us know that radiation comes in discrete amounts called quanta. This gives us a panoramic view of the whole Quantum World – behavior of atoms and their particles. From the Quantum Theory, Werner Heisenberg developed the theory of Quantum Mechanics based on the Uncertainty Principle. Mechanics is the term used for the study of the motions in the Quantum World. But Heisenberg's Quantum Mechanics is equally significant because it has shown us the fundamental difference between our World and the Quantum World.

This fundamental difference is that one can only measure or probe a particle by using a comparable quantity, a quantum of light. A quantum of light will disturb its position and interrupt its velocity. Continuous efforts to determine the position and velocity of a particle mean that the more accurately the position is determined, the less accurately the velocity is determined. Velocity and position cannot be independently determined. Therefore, velocity and position are given a quantum position, a blending of the two. The Theory does not predict a definite result of an observation, but a possible result. Quantum Mechanics had grave consequences for Determinism. If one cannot measure the present velocity and position of a particle, one cannot predict its future velocity and position. If one cannot measure the present state of the Universe, one cannot predict its state at any other time.

The beauty of the Theory is that we can apply it to everyday life. It means we must use our discretion in the daily circumstances of life. We should act within the limits set by Quantum Mechanics. What this means is that we should always give someone else the benefit of the doubt, but we should not go to an extreme to do it: for to do so would not be within the limit allowed by Quantum Mechanics. We still need to use our discretion when we give someone else the benefit of the doubt
.

The Theory of Relativity

The Theory of Relativity is synonymous with the name Albert Einstein (1879-1955). Einstein gained world fame for the Theory due in part to his scientific brilliance expressed in one aspect of the Theory and in part to the controversial nature of another aspect. There are three aspects to the Theory as seen in Pre-Relativity, Special Relativity, and General Relativity.

Pre-Relativity Physics is summed up thus: "In the first place it is assumed that one can move an ideal rigid body in an arbitrary manner. In the second place, it is assumed that behavior of ideal rigid bodies towards orientation is independent of the material of the bodies and their changes of position, in the sense that if two intervals can once be brought into coincidence, they can always and everywhere be brought into coincidence." The Meaning of Relativity published by Princeton University Press, N.J., Fifth Ed., p 4-5

Pre-Relativity physics could be seen as a subtle attempt aimed at the very foundation of Newton Mechanics. Einstein did not believe in Gravity in the Newtonian sense. The motions of two bodies are independent of the contents of their matter.

Special Relativity explains the equation of energy and mass and limits the speed of all material objects. No material objects can acquire the speed of light. The Michelson-Morley Experiment of 1887 was critical to the Theory. In fact, it laid the foundation for the Theory. Michelson and Morley, in their experiment, used the Earth as an example. If the Earth is moving through the Ether, its motion should cause an Ether breeze and a beam of light traveling against the Ether breeze should have a slower velocity than a beam of light traveling across it, they reasoned. They invented a device called the Interferometer to conduct their experiment. It showed that there was no difference in the speeds of the two beams of light. However, these results were inconclusive because they were challenged by Irish scientist, George Francis Fitz-Gerald in 1892.

A year later Fitz-Gerald gained the support of Dutch scientist Hendricks Anton Lorentz. This joint response to the results of the Michelson-Morley Experiment came to be known as the Fitz-Gerald Lorentz Contraction. The Fitz-Gerald Lorentz Contraction proposed the possibility of the arm of the Interferometer pointing into the Ether breeze being shortened, reducing the distance the beam of light would travel. This decrease in distance would be responsible for its faster speed and would compensate for the slower speed of light of the longer arm traveling against the Ether breeze. Both beams of light would, therefore, be recorded at the same speed. The important thing was that there was no Ether Breeze.

With the appearance of Special Relativity 1905, all inconclusiveness was removed from the results of the Michelson-Morley Experiment. So the results of the experiment became the premise of Special Relativity. The speed of light is constant and should be the same to all moving observers irrespective of their speeds of motions. Special Relativity introduces a new interpretation of time, and space. Special Relativity shows how time is related to space. This relativity is obvious from the fact of an event that occurred at some point in space. For this particular event there are three observers at three different locations in space. Each observer has a perfectly accurate clock, but owing to each one's distance in space from the point of this event, each one's clock will record different times and yet be correct.

The determining factors in the relativity of time and space are the speed of light and distance. Light travels at a speed of 186,000 miles per second. An event occurred in space. A ray of light is beamed from the event to three moving observers at three different locations from the event; 93,000,000 miles, 46,500,000 and 31,000,000 miles. The clock of one observer will show a time of 8 minutes; the clock of another will show a time of 4 minutes, and the clock of the third observer will show a time of 2 and two thirds minutes. These three recorded times are correct but they have not changed the absolute nature of time because the determining factor of the speed of light is Earth's standard of time. General Relativity explains the motions of Celestial Bodies on the principle that Space is warped by the masses of matter and that the Celestial Bodies continue orbiting in the warped spaces.

Response: This aspect of Relativity corresponds to Newton's first law of Motion. The exception is that Einstein makes these orbits independent of their magnetic properties, but solely dependent on the warped spaces. Absent from both theories are two important factors: the weightlessness of outer space and the Magnetic-field of the Universe. Without the weightlessness of outer space, all motions would be impossible. The Magnetic-field of the Universe provides the microphysical entities and processes absent from Newton Mechanics. None of these two theories explains initial orbits. Initial orbits began during the formative state of these bodies and thus continue. If one puts an object into space at a certain speed, it will continue orbiting the Earth at that same speed.

Astronomers have seen more than 40% of the stars orbiting one another.. How could this be if there were no gravitational attraction amongst the stars? The question then is, Which of the two theories is closer to reality? Of course, Newton Mechanics is the better theory.

Chapter 4: The Mind of God, a Schedule of Laws Descriptive of the Universe.

Knowing what we know about the Universe, it is correct to say the Universe was preplanned. This chapter shows that all the laws of science were developed in the mind of God and began to materialize the moment He gave the command of Creation. It shows how an imaginary Universe became a natural reality through such laws as the Laws of Weightlessness, Gravity and Motion and the Laws of Variation.

Weightlessness

The invisible matter became the foundation of the Universe by causing the condition of weightlessness of outer space. It is like ships in the oceans. The volume of water occupied by a ship is heavier than the ship, so it floats on. On this principle numberless ships sail the oceans. Thus all bodies in space are in a state of weightlessness. This fact can be illustrated from our daily experience. Each time we look at the skies we see volumes of clouds moving in all directions. We know that this has been possible because of the air pressure of the Atmosphere. We do not see the Atmosphere but we know it is there and causes the clouds to move.

The Laws of Gravity and Motion

The Laws of Gravity and Motion revealed the mind of God before He created the Universe because these laws did not just impose themselves upon the Universe. These laws were the focus of attention elsewhere. The illustration of the magnet and the wire shows plainly the effect of Gravity of one body on another.

One can deduce that the wire and the magnet have much in common and that the basic difference between the two is that the magnet has more magnetism than the wire. The two principles in this mysterious happening are the similarity in matter and the magnetic fields. The action of Gravity must work by these two principles. Matter is basically composed of electromagnetic properties. This is particularly obvious of Celestial Bodies (stars). Radiation of stars is 100% electromagnetic in nature. This radiation creates and sustains the Magnetic-field of the Universe. The Magnetic-field of the Universe works in conjunction with the magnetic-fields of the attracting mass and the attracted mass. The Universal law of the Conservation of Energy (energy cannot be created or destroyed) dictates that the radiation of stars must remain in the Universe in some form.

One cannot separate the Magnetic-field of the Universe from the attraction of Gravity. They are critical to the motions in the Universe. The evolutionists will say if that be the case the Universe would have collapsed many times over. This would be true if the Universe had accidentally evolved. And still this would not be an over-night event because the attraction of Gravity is balanced by the velocity of motions.

Newton Mechanics did not attempt to explain how rotations and orbits were established. It explains what happens after the mechanism of the Universe was switched on. While Newton Mechanics did not show how velocities of motions and orbits were established, it graphically shows that velocity of motion balances the attraction of Gravity. If the planets were not moving as they do, they would have eventually crashed into the mass of the Sun.

Laws of Variations

The Laws of Variation do not explain why matter should be in different forms or kinds. Different forms of matter were by Divine purpose. However, in each different form of matter there are variations. These laws explain those variations. In the formation of the elements, the stars and the planets, the laws of variations apply. The human race is not exempt nor is any other thing in nature, in the microcosmic world as well as in the macrocosmic.
Note the following 3 laws:

1).The nature and volume of matter determine the variations in that particular form of matter. The greater the volume of matter, the more diverse are the variations. But to a great extent, this depends on its nature. A chalk coal when falling from a certain height onto a hard surface will have broken into several pieces and each will have varied in size and shape. If a larger chalk coal is falling from a higher height, it will have broken into many more pieces, each piece varying much more from the other. If that larger chalk coal was more dense or harder than the smaller chalk coal, it might have broken into the same number of pieces, or even fewer number of pieces, depending on its density or hardness.

Even in the Cosmic Background Temperature (the temperature of outer space of 454 F below zero), there are variations due to the immensity of the Universe. It should not surprise anyone when scientists reported in April, 1992 that they found variations in the Cosmic Background Temperature. What should have been the surprise was the fact that they had not thought of that before. And even more astonishing is their interpretation of these variations as the "relics of the Big Bang".

2). The process of formation of a particular form of matter will determine the variations of that form of matter. All kinds of coal are made from the original matter of dead plants and animal matter. But different stages in the process produce coal variations. Peat is mainly a mixture of carbon and water with other minerals. Brown Coal is a further process of Peat. Soft Coal is a further process of Brown Coal. When all the water is processed out of coal, the final result is hard carbon.

3). The life span of a particular kind of matter will determine variations. (a) At the beginning of the life span of a particular kind of matter, variations are less diverse. (b) At the middle of the life span of a particular kind of matter variations are more diverse. (c) At the ending of the life span of a particular kind of matter variations are most diverse. The clearest evidence of the laws of Variations can be seen in the human race.

Chapter 6: Motions of Bodies in Space

Astronomers, indicate that 20% of the visible stars orbit each other. In many Science Books it is emphatically stated that the stars are directly orbiting around the core of the Galaxy of the Milky Way. This statement directly contradicts scientific observation. They could not be directly orbiting each other and simultaneously orbiting the core of the Galaxy. This idea is based on E Mach's Theory. In his Book The Meaning of Relativity, Einstein was careful to emphasize that E. Mach's attempt to eliminate space as an active cause in the system of mechanics, by suggesting that a material particle does not move in un-accelerated motion relative to space or uniform motion relative to space, but relative to the center of all other masses in the Universe. Einstein was also careful to point out that E. Mach had failed in his attempt. It is clear that the orbit of the Sun as explained by scientists is based on Mach's condemned theory. The Meaning of Relativity, Princeton University Press, 1984 edition, page 56

Applying Mach's theory to the motions in the galaxy of the Milky Way means that the Sun would not orbit the closest center of Gravity, but instead would be orbiting around the center of the Milky Way. This is only part of the fact of the Sun's orbit and not the greater part of that fact. Indirectly the Sun orbits around the center of the Galaxy but not directly. The Gravity of a mas is what holds its matter together. This means indirect orbits of the stars. To say each star orbits the core directly, is preposterous.

Let us now picture the gravitational attraction to the Milky Way which is the collective gravitational attraction of the individual stars. This is what holds the Galaxy together. When we apply E. Mach's Theory to the Galaxy, it means that the motions of the stars are inclined towards its center, but the theory does not recognize local centers of Gravity within the Galaxy. That is why I have said the Theory represents a part of a fact and not the greater part of that fact.

Since astronomers have reported star systems in which two stars orbit each other and systems in which three or more stars orbit each other, local gravitational centers within the Galaxy have been established. We can visualize the gravitational center of the Galaxy as a core of stars, quite possible the most massive stars orbiting each other. A number of other stars may be directly orbiting around each other. Other stars may be directly orbiting around this nucleus. If the information about the Black Hole at the center of the Galaxy is accurate, this construct is also correct

Rotation of the Galaxy

Scientists suggest that in hundreds of millions of years the stars in the Galaxy will eventually orbit around its center depending on their respective distances from it. It is very convenient to say this, as it is the popular thing. My idea of the rotation of the galaxy is simple. Stars have two basic motions. They rotate and move in a circular line within various groups. Since the stars in the Galaxy sustain these basic motions, it must be understood that the Galaxy is rotating. This simple and logical way of explaining the rotation of the Galaxy is not what is meant by the following authors:

We live in a galaxy that is about a hundred thousand light years across and is slowly rotating; the stars in its spiral arms orbit around its center about once every several hundred million years. Our Sun is just an ordinary, average-sized yellow star near the inner edge of one of its spiral arms. A Brief History of Time op. cit. p.37

It has also been found that the Galaxy is rotating around its center; the Sun takes 225,000,000 years to complete one revolution period which is officially called the cosmic year. The Amateur Astronomers' Glossary by Patrick More p.50. Rotation of this galaxy will, therefore, in the course of some tens of millions of years, cause every stellar association to spread out in the direction of rotation and assume an oblong shape. Atlas of the Universe op. cit p.99.

This scientific explanation of the rotation of the Galaxy cannot, to say the least, be supported by any scientific evidence. It is based on E Mach's theory which had as its premise, the elimination of space as an active agent in the system of mechanics (Newton's Theory of Gravity and Motion).

The rotation of the Galaxy, as explained by scientists and the disintegration of its constellations, by reason of its rotation, have been swiftly rejected. The Theory of the Disintegration of Star Constellations has been rejected on the grounds that it conflicts with the emphatic statement of the Creator regarding the stability of star constellations: "Canst thou bind the sweet influences of Pleiades, or loose the bands of Orion?" (Job 38:31, 32).The theories rejected in this chapter and elsewhere have not been done out of disrespect for the devotion of scientists to provide us a better understanding of the Universe. These theories have been rejected because some of them are self-contradictory, and as such are an obstacle to a better understanding of the Universe. If we must have a better understanding of the Universe, as is our aim in this book, the obstacles to such understanding must be eliminated.

Reconciling the Theory of the Rotation of the Galaxy with the Theory of the Expanding Universe

Scientists explain the Universe is expanding; galaxies are racing from each other. At the same time they explain galaxies are rotating. Can the two be reconciled? Let us give this our attention. In our review and response to scientific theories in chapter 3, the Expanding Universe was discussed. The substance of the Theory is that galaxies at the outer reaches of the Universe are receding from each other at incredible speeds; the farther they are away, the faster they are moving. This observation forms the foundation of the Theory. Part of our response to the Theory there and then was that if one were to view the center of the Universe from its outer edges, one would have seen the galaxies at the center spreading farther and farther apart with similar velocities. So how could one reconcile the Theory of the Expanding Universe with the Rotation of the Galaxy?

The expansion of the Universe would have very serious consequences for the stability of the Universe. First of all, it would have caused the constellation of stars to spread out and not the so called rotation of the Galaxy as some scientists suggest. Orbits would be affected by any meaningful expansion. Space represented dark invisible matter. This created condition of weightlessness. The idea is that a galaxy has more invisible matter than visible matter. This supports the weightless condition of a galaxy. If a galaxy keeps spreading apart from others, very soon it would collapse because it would be losing the over-all support of invisible matter. It is like the foundation of a house which keeps rending apart. Very soon parts of it will begin collapsing. Before long the whole house will collapse. It does not really make sense in talking about the rotation of a galaxy when the conditions of rotation are absent. Does it? Scientists should be more thoughtful before declaring their hypotheses

A reconciliation between the two theories is just not possible. An assumption or theory can describe or even predict a reality but an assumption and a reality are not always the same thing. At times an assumption can be remote from reality beyond imagination. And we here have a classic case before us. Since one cannot reconcile the Theory of the Expanding Universe with the Theory of the Rotation of the Galaxy, the reality is that neither the Universe is expanding nor the galaxies are rotating the way scientists explain. One obstacle to a true scientific description of the Universe is the self-contradictions of the scientists. A true scientific description means the elimination of all such contradictions.

Chapter 7: The Economy of Cosmic Radiation and Recycling

This chapter shows the importance and the interdependence between Cosmic Radiation and Cosmic Recycling in the system of the Universe, hence the name, the Cosmic Compound. This chapter also shows the importance of the Sun's radiation to the sustenance of life on Earth, that the greater percentage which goes into outer space is not wasted, and that the radiation of the stars is not wasted. Based on the Universal Law of Energy Conservation, the radiation of the stars remains in the Universe in some form or another. In particular, most of this radiation is collected into the Magnetic field of the Universe through which Gravity works. There is also a special emphasis placed of the Radiation Cycle.

For some time, before I began writing this book, I could not escape the idea that there was something more about star radiation, some secret that the scientists have not yet discovered. Then the idea of a radiation cycle impressed itself upon my mind. The idea is that the Sun's three layers of gases interact to produce a Radiation Cycle. This idea stands in direct contrast to the scientific Theory that within 5,000,000,000 years, the Sun will have exhausted its nuclear fuel. The Sun was created to manufacture its nuclear fuel which it has been doing since Creation. The Sun's inner core sustains a temperature of 20,000,000 degrees Centigrade. This temperature is responsible for the abundance of the Sun's Hydrogen.

We must now inquire as to how the Sun's radiation cycle is sustained. There are three known layers of gas of the Sun: the Photosphere, the shinning disk; the Chromosphere lying above; and the Corona lying above the Chromosphere. These layers of gas range in temperature from 6,000 degrees Centigrade to 1,000,000 degrees Centigrade with depths from 10,000 to 1,000,000 miles. The Photosphere is a mirror of what is happening in the Sun's Radiation

Cycle. In the Photosphere Hydrogen Atoms fuse to produce Helium, causing radiant energy. All three layers of gas are held together by the Sun's Gravity from which they can never escape; all that can escape is its radiant energy. With the Sun's Gravity and with its inner temperature of twenty million degrees Centigrade, its radiation cycle is sustained.

We are no strangers to Gravity; we know what it can do and we know that the Sun's gravity is many times greater than that of the Earth. Therefore, the Sun's Gravity and its gases create its radiation cycle, not identical to the Earth's but similar. This radiation cycle could last forever, bearing in mind the Universe was created to last forever. Our friends, the scientists, tell us that most of the stars like the Sun will have exhausted their nuclear fuel within 5,000,000,000 years; yet they have failed to make any prediction of the Earth's water cycle, which logically would be more precarious than the Sun's radiation cycle.

Many of us would like to remember most of what we have read in a book, but so few of us have a good memory. This review is an emphasis on the subjects of this book; and of such, is intended to facilitate memory.

Concluding Statement 205

Reflecting on the things herein said, we can conclude that enough has been said to give a graphic scientific description of the Universe so that the average child is able to have a general understanding. Modern science has alleviated some of the burden of the curiosity, but has failed to give contemporary man the description of the Universe he has so eagerly anticipated. It has failed chiefly for the fundamental reason that it has ignored or overlooked that indispensable factor in all knowledge—GOD.

This is glaringly obvious in The Evolution Theory of the Universe and the origin of life. The Big Bang Theory is succinct and incomplete, not only from a Theological point of view but also from a scientific point of view. This is so because it has failed to explain how the matter of the Primeval Atom of 100 million miles in diameter originated. It has also failed to address in a fundamental way the question of space and the weightlessness in outer space.

With regards to the evolution of life, the Theory has failed in the most critical aspect— the explanation of the millions of different species in the animal kingdom. And still the human species has presented more complexities for the Theory because it cannot explain the moral and spiritual nature of man.

If the origin of the Universe has not been intelligently explained by modern science, neither can it explain the purpose of the Universe. This is also true of the origin of human life. The failure of science to answer these fundamental questions is adequate proof that it cannot provide a true and complete description of the Universe. It is like an amazing paradox the way science describes nature. Science does not believe in miracles but its description of nature amounts to a series of scientific miracles.

A true and complete scientific description of the Universe necessitated a true Theological Postulate. The Theological Postulate herein expounded has met this requirement. At its center is an Eternal, Intelligent, Omnipotent, and Infinite Being. Since this Postulate is founded on the Bible, it must be authentic and is the kind that meets the expectation of religion and the challenges of an infinite Universe. The challenges from an infinite Universe have less to do with modern discovery and more to do with the logical deductions from these discoveries.

How can one deduce from these discoveries the most fundamental questions about the Universe? One of the most of all fundamental questions has to do with the relation of the invisible matter with the visible matter of the Universe. And although the Big Bang theory implies such relation that the invisible matter far outweighs the visible matter; scientists have failed to address the matter in a definitive way. There could be nothing more fundamental to our understanding of the Universe. For some strange reason, the science books have failed to provide this information. Certainly, the lack of this knowledge makes the Universe more mysterious than what it really is. On the other hand, it shows that the Universe could not have accidentally evolved. The Theo-Cosmos Theory postulated in Chapter 2 has sufficiently addressed the foundation of the Universe. A preponderance of the evidence is obvious from the Cosmic Microwave Radiation Background, radiation of stars, and the weightlessness of outer space.

For some time the atom was once thought to be the smallest unit of matter and thus completely dense. But microscopic examination over a period of time not only reveals that the atom is made of smaller particles, but how many and how it is constructed. This revolutionary discovery should tell us that in order for the masses of matter in the Universe to behave the way they do, some other factor had to be present.

We have learned much from science about the laws of Gravity and Motion and yet failed to identify the proper orbit of the Sun. Herein, we take exception to the general rule, addressing the difficult scientific questions and not seeking to avoid them. We have addressed the question of space, the origin of matter, motions and orbits, Cosmic Recycling, the 23 degree tilted angle of the Earth, the

Magnetic-field of the Universe, the weightlessness of outer space and the foundation of the Universe.

We were quick to respond to some of the better known scientific theories. We responded with clarity and balance. We unequivocally rejected the scientific Theory that the Sun is orbiting the center of the Milky Way. And, instead, with the help of revealed science, we were able to chart the Sun's orbit. We have objected to the scientific deduction that over a period of time the rotation of the Galaxy will have caused the disintegration of its constellations of stars. This Theory directly contradicts the scientific information in the Bible regarding the stability of star constellations. [Job 38:31, 32]

The scientific observations of the infinity of the Universe have been warmly welcomed. These observations have confirmed the scientific pronouncements of the Bible. "The hosts of heaven cannot be numbered" [Jeremiah 33:22]. Modern discoveries of thousands of galaxies like the Milky Way give cause for the theory of the expanding Universe, but give no justification for that Theory. Discovering the immensity of the Universe does not necessarily mean the Universe is expanding. But why is the Universe the way it is and why is the Universe here? First, the Universe by design must reflect the infinity of the Creator. And there must be a purpose for the whole realm of nature. Since the Universe reflects God's infinity, it must have an eternal purpose that will never change.

In the first place, its purpose is to serve God's own honor. In the second place, the Universe is the medium by which and through which God shares His honor and His life with His creatures. Life thus becomes an eternal privilege. The most complicated mathematical equations cannot answer the basic questions of the Universe. Thus the human thirst for knowledge would have remained unquenched.

In closing our reflection on some of the things said in this book, we feel obligated to at least acknowledge some of the possibilities. Are there billions of systems like our Solar System? Are there billions of civilizations like ours? We do not know with certainty these answers.

However, we do know that if we were the only civilization in this Universe, the purpose of Creation would still have been served. God had to create an infinite universe because only by doing so would men and angels know and appreciate the true meaning of His Infinity. Knowing this, men and angels shall join hands together in the Praise and Worship of Him Who is eternally worthy.

And when I stand before my Creator, He may look on this scientific description of the Universe and point out some things to me or mistakes I have made. Then I shall say to Him, I inherited a sinful nature, because of this my work was not perfect. Then he will look at me and say to me, Enter into My Kingdom not because of your scientific description of the Universe but because you accepted Jesus My Son as your personal Savior and Lord. My heart shall then overflow with Praise and Gratitude. Then I shall say, O Father, my Earthly sorrows cannot justify the privilege of one moment in Your Eternal Presence.

This book does not encroach on the incomprehensible greatness of the Eternal Infinite God; rather, it seeks to glorify His greatness.

May God Bless Mankind And Grant Us All A Place In His Eternal Kingdom.

A

Accidental Evolution of the Universe: The observable Universe of 200 to 500 billion galaxies evolved accidentally, 16, 25.

Antediluvian World: The people who were destroyed by the Flood in Noah's day, 20, 23

Attributes of God: His qualities, characteristics, 21-22.

Arcturus : A constellation of stars, 39, 127.

Archeozoic Era: The first of six periods of time the geologists say the Earth was developing, 47

Atom: The smallest unit of matter made of a nucleus of neutrons and protons and orbiting electrons 103, 104

Atomic number: The number of protons in an atom, 107.

Atomic mass or weight: The number of protons and neutrons of an atom, 107.

Alpha Decay: One of three levels of radioactivity, 108

Atmosphere: The whole blanket of air surrounding the Earth, extending approximately 400 miles above, 109

Asteroid: A small island of metal and rock orbiting the Sun, 115.

Alpha Centauri: The three star constellation nearest the Sun, 129.

B

Bible: The text book of the Christian Religion, the inspired and revealed Word of God, 17-18.

Biological Evolution: Shortly after the first living cell was produced by the Chemical Evolution of millions of years, that living cell multiplied and filled the oceans according to the theory of Evolution, 50.

Bond: The force that holds the molecules of matter together, 107.

Beta Decay: One of three levels or radioactivity, 108.

Black Holes: Great voids in space, the collapse of a star under its Gravity in which its radiation is overcome by its gravity, 119-120.

C
Congruity:
One of the Theological arguments for the existence of God is the argument from CONGRUITY. It states that a theory which explains all the related facts in a case must be correct. The belief in the existence of an Omnipotent and Eternal Being explains the facts of man's mental, moral and spiritual nature as well as all the facts of the material phenomena of our Universe, 15.

Cosmos: The Universe, 15

Cosmology: The study of the arrangement of the Universe, 15.

Character of God: The sum total of His activities and actions in relation to His holy nature, 22-23.

Creation: The works of God, 24, 26.

Contradistinctions: The two great contrasting distinctions in the formation of the Universe; the Miraculous Aspect and the Scientific Aspect, 25, 26.

Cosmological Constant: A theory proposed by Albert Einstein, repulsive built in force in nature to balance the attraction of Gravity, 55, 73.

Cold invisible matter: Most of the matter in the Universe, causing the weightlessness in outer space, 39, 140.

Cosmic Cream: The clouds of dust and gas which formed and floated in the cold invisible matter at the beginning of the Universe, 39

Carbon: One of the chemical properties of life, 49, 50.

Calcium: One of the chemical properties of life, 49,50.

Cenozoic Era: The fifth of the six eras the geologists say the Earth was developing, 48.

Chemical Evolution of life: The chemical elements provided by the interactions between the early atmosphere and the oceans that were the precondition for the Biological Evolution, 50-51.

Cosmic Umbrella: The planets and all the stars within the neighborhood of the Solar System have so warped space that a sky above is the result, 58, 59.

Centripetal Force: One of the two great forces in the Universe. It makes motions inclined inward, 62.

 Centrifugal Force: One of the two great forces in the Universe. It makes motions move outward, 62.

These two great forces are inwrought in the very fabric of the Universe.

Compound: Matter made of two or more different kinds of elements or several elements, 107,

Centigrade: The alternate scale to Fahrenheit in measuring, 98, 114.

Corona: The layer of gas lying above the Chromosphere of the Sun about a million miles in depth, 114, 142.

Chromosphere: The layer of gas lying above the Photosphere of the Sun about 10,000 miles in depth, 114, 142.

Comets: A mass of frozen particles and gases about a mile in diameter orbiting the Sun, 115.

Cosmic Clouds: Gigantic clouds of dust and gas in outer space caused by the interactions between cosmic radiation and cold invisible matter, 139, 165

Cosmic Radiation: The radiation from all radiant bodies in the Universe, 138.

Cosmic Recycling: The recycling of matter including radiation in various forms in the Universe, 135,140.

D

The Doppler Effect is applicable to sound as well as radiation. In the case of sound, the frequency is higher moving towards its source, depending on the speed at which one moves. The frequency or sound is lower, moving away from the source, depending also on the speed, 56.

Determinism: The theory proposed by French scientist, La Place, that a supreme intelligence possessing a knowledge of the Newtonian laws of nature and a knowledge of the positions and velocities of all particles in the Universe at any moment, could determine the state of the Universe at any other time, 77, 78.

E

Ether:

The Theory of the Ether, in its essential nature, was that an undetermined substance permeated the Universe solely for the purpose of propagating light. Light needed a medium through which to manifest. The Ether was considered to be of absolute position, motionless. It should determine the motions of all bodies in the Universe, 71.

Ether Breeze: In Mitchel-Morley Experiment no Ether Breeze was discovered, thus confirming the absence of the Ether, 71.

Earth's Crust: The upper layer of the Earth of a depth of 20 miles, 111.

Earth's Magnetic-field: Air molecules in the Ionosphere electrically, charged, forming an electric blanket around the Earth, 89.

Ecliptic: The apparently yearly path of the Sun among the twelve constellations of the Zodiac, 148-149.

Electromagnetic Force: The second of the four elementary forces of mater; the second weakest, 105

Electron: Negatively charged particle orbiting the nucleus of an atom, 105.

Element: Matter made of one kind of atom, 107

Elementary Forces: The forces arising from the interactions between the particles of matter, 105.

Expanding Universe Theory: The Universe is expanding but just at the critical rate to balance the attraction of gravity suggested by scientists, 55, 56.

Exosphere: One of the four layers of air in the Atmosphere, 109,110

Equation of Energy and Mass: Critical to Special Relativity is its equation of energy and mass. The Equation is spelled out as $E=MC2$, wherein E is energy, M is mass, and C is the speed of light. In practical terms, energy can be converted into mass and mass can be converted into energy. Energy equals mass multiplied by the speed of light squared. This equation tells you what potential energy is in a particular form of mass. Therefore, the energy a mass acquires through its motion increases its size. In the final analysis, when a material object is travelling close to the speed of light, its size doubles. However, it will never acquire the speed of light. Therefore all material objects are limited to the speed of light, 72.

F

Foundation of the Universe: The **cold invisible matter** far exceeded the clouds of dust. The clouds of dust then floated on. With the far greater amount of **cold invisible matter,** the condition of weightlessness was created — the foundation of the Universe was established, 39.

Fitz-Gerald Lorentz Contraction: The Theory proposed 1892 by scientists George Francis Fitzgerald and Hendricks Anton Lorentz. It challenged the findings of the Michelson-Morley Experiment, 71-72.

Fine-tuning of the Universe: The fully developed Magnetic-field of the Universe and the established motions (rotations and orbits) of bodies in space, through the laws of Gravity and Motion, 68139.

G

Gravitation Waves: loss of energy by motions of bodies in space. Over a period of time, millions of years, a body in space loses energy in motions. This loss affects velocity in motion, 60, 61.

Galaxy: A star system of 100 billion stars, 25

Gamma Decay: One of three levels of radioactivity, 108.

Geocentric Theory: The theory that the Earth was once the center of the Universe, 11 (in introduction)

Gravitational Force: One of four elementary forces of matter, the weakest, 105,

Gravity: The attraction of one body on another, 65, 66.

H

Hot Big Bang Theory: The accidental beginning of the Universe, 31, 32, 33.

The Hubble Discovery in the 1920's showed that the galaxies at the outer reaches of the Universe were moving away and the farther away, the faster they were moving, 56.

Hydrogen: One of the chemical properties of life; the most abundant element in nature, 97.

Hydrosphere: the liquid part of the Earth; 110.

.Half Life of an Element: The period of time for an element to form a more stable form of matter is called the half-life of that element. One gram of radium (1/32 of an ounce) will lose half its mass in about 1700 years and settle down to a more stable form. One gram of Uranium, U 238 will lose some of its mass in 5 billion years and settle down to the more stable form of lead, 108.

I

Infinity of God: By the infinity of God is meant that He is free from all limitations. In the first place, He is not bound to time. In the second place, He is not bound to space, 22, 174.

Immanence of God: God is present in nature, overseeing the Universe, 29

Intellectual endowment of man: This endowment allows man to formulate the Theo-Cosmos Theory of the formation of the Universe, 25, 30, 171.

Infinity of the Universe: The Universe cannot be measured by scientific means, man can only have an idea of its size, 22.

Inner Core: One of four layers of rock comprising the solid part of the Earth having a depth of 800 miles, 111..

Inner Mantle: One of the four layers of rock comprising the solid part of the Earth having a depth of 1300 miles, 111.

Interferometer: The device used by Albert Michelson and Edward Morley in their 1887 experiment of the Ether, 71, 72.

Ionosphere: One of the four layers of air in the Atmosphere. It lies between the Stratosphere and the Exosphere. It begins above the Stratosphere and extends several hundred miles above. In it lies the Earth's magnetic field. Here the air molecules become ionized electrically charged, forming an electrical blanket around the Earth. The Ionosphere facilitates radio transmission. Without it there could be no radio transmission, 109.

Iron: One of the chemical properties of life, 49..

L

Law of Apperception: The method by which one teaches something new by starting from that which is already known, 70-.

Laws of Gravity and Motion: Principles explaining gravity and motion, 65, 66..

Laws of the Elements: Principles explaining the formation of the elements, 95, 96.

Laws of Variations: Principles explaining the variations in a particular form of matter, 98, 99.

Laws of the Horizontal position of the Universe: Principle explaining the horizontal position of the universe, the Galactic plane, 100-101

Laws of Distance: Principles explaining distance in relation to particles of matter and bodies in space, 101- 102.

Laws of Weightlessness: Principles explaining weightlessness in outer space, 84-85.

Lithosphere: The solid part of the Earth, 111.

M

Microwave Radiation: The temperature of outer space of 454 degrees below zero; electromagnetic waves of high frequency, 59.

Mazzaroth: The name of a constellation of stars, 39, 127.

Magnetic-fields local: the magnetic fields of two bodies associated in a gravitational interaction, 68, 89.

Magnetic field of the Universe: Cosmic radiation is collected into the

Magnetic-field of the Universe through which gravity works in conjunction with the local magnetic field, 68, 89.

Mesozoic Era: The first of six eras in the development of the Earth suggested by the geologists, 48.

Microcosmic and macro-cosmic: The two extremes in matter, very small and very large, 31, 101

Michelson-Morley Experiment: The experiment conducted by American scientists Albert Michelson and Edward Morley in 1887 which abolished the theory of the Ether and showed that the speed of light was constant irrespective of the speed of motion of an observer, 71, 72,193.

Molecule: A group of two or more atoms, 50,107.

The Milky Way: The Milky Way is one among billions of galaxies; it our Home Galaxy of 100,000,000,000 stars. If it were viewed from one of those receding galaxies, there would certainly be a red shift, even though scientists inform us the Milky Way is rotating on its axis. Looking from one of these receding galaxies, astronomers would see the galaxies at the center of the Universe spreading apart, though they are not now thus seen, 25, 57.

N

Natural Selection: Charles Darwin's theory of self-preservation in the animal kingdom, 51, 52.

Neutron: Uncharged particle of the nucleus of the atom, 104.

Newton Mechanics: Sir Isaac Newton's theory of gravity and motion, 65, 69.

Nitrogen: One of the chemical properties of life, 49.

Nova: An ordinary star explosion, 145.

Nuclear energy: The energy resulting from nuclear fission or splitting the atom. Two German scientists, Otto Hahn and Fritz Streseman were the first to split the atom in 1939. They bombarded Uranium 235 with high speed neutrons, 106.

O

Observable Universe: 200 to 500 billion galaxies each having 100 billion stars, 25.

Outer Core: One of four layers of rock comprising the solid part of the Earth having depth of 1240 miles, 111

Outer Mantle: One of four layers of rock comprising the solid part of the Earth having a depth of 600 miles, 111

Oxygen: One of the chemical properties of life, 49.

Ozone Gas: A layer of gas in the Stratosphere which converts the energy from the Sun into heat, 109.

Orbital velocity of the Sun: At an orbital velocity of 135 miles per second, the Sun travels 4,245,696,000 miles in one year. A light year is 6,000,000,000,000 miles, the distance light travels in one year. At an orbital velocity of 135 miles per second, the Sun would need more than 1200 years to travel the distance of a light year. The Sun needs more than 4,800 years to reach the nearest constellation, 159

P

Principle of Center Formation: The Principle by which stars are formed in a galaxy, 40.

Paleozoic Era: The third of six eras geologists say the Earth was developing, 47.

Perihelion: The nearest point where the planets, come to the Sun, 76

Photosphere: The shining disk of the Sun, 114.

Planetary nebula: A mass of spherical cloud illuminated by a central star, 118.

Positron: An anti-particle of the electron,105.

Primeval Atom: The original matter of 100 million miles in diameter which scientists say exploded causing the birth of the Universe, 32.

Primordial Soup: The theory that the radiation of the Sun and the lightning converted the simple gases in the upper layers of the oceans into complex carbon based molecules, 50.

Prism: A glass triangle that collects light rays forming the spectrum of

Rotation of the Galaxy: The rotations and orbits of stars in the Galaxy, 130-132.

S

Solar System: The Sun and nine planets, 113.

Spectroscope: An instrument used with the telescope to study the light from stars, 122.

Static Universe: A non-expanding Universe, 56.

Stratosphere: One of four layers of air in the Atmosphere, extending about 50 miles above the Earth, 109.

Strong Nuclear Force: The fourth and strongest elementary force of matter, 105.

Supernova: An extraordinary star explosion, 145.

T

Theo-Cosmos: A reconciliation between, science and Theology, 17.

Theologian: One who studies about God and His relation to the Universe, 15, 16.

Theology: The study of God and His relationship to the Universe, 15.

The Universal Law of the Conservation of Energy: Energy can never destroyed, can only be transferred into other forms, 35, 60.

Thermocouple: An instrument used with the telescope to measure the temperature of stars, 122.

Transcendence of God: God being eternal is before time, 22.

Troposphere: One of four layers of air in the atmosphere extending about 10 miles above the Earth, 109, 110.

W

Weak nuclear force: the third of the four elementary force of matter, 105

World War 2, 106.

Byron Preiss: The Universe, Batman Books Inc., 1987 Ed.

The Encyclopedia of Astronomy and Space, Thomas Y Crowell, NY

Fred Hoyle: Galaxies, Nuclei, and Quasars, Harper and Row, NY

The Encyclopedia of Philosophy: Crowell Collier and Mac Millian Inc. *Ed.*

Richard Laurence: The Book of Enoch, Wizard Bookshelf, 1976

Charles Darwin: The Decent of Man, The University of Oxford Press

Philip J. Gearing and Graig Conrad: Natural Science; Stech Vaughn Co, Austin, Texas

Basic Science: Cambridge Book Company, New York

Joachim Ekrutt: Stars and Planets, Barrons

Stephen Hawking: A Brief History of Time, Batsman Books, 1990 Ed.

Henry C. Thiessen: Introductory Lectures in Systematic Theology, Wm. Publishing Company, 1963 Ed.

Hazen and Trefil: Science matters, Doubleday, 1991 Ed.

Atlas of the Universe: Thomas Nelson and Sons, 1961 Ed.

Albert Einstein: The Meaning of Relativity, Princeton University, Fifth Ed.